T0213599

Lecture Notes in Computer Science　　10511

Commenced Publication in 1973
Founding and Former Series Editors:
Gerhard Goos, Juris Hartmanis, and Jan van Leeuwen

Editorial Board

David Hutchison
　Lancaster University, Lancaster, UK
Takeo Kanade
　Carnegie Mellon University, Pittsburgh, PA, USA
Josef Kittler
　University of Surrey, Guildford, UK
Jon M. Kleinberg
　Cornell University, Ithaca, NY, USA
Friedemann Mattern
　ETH Zurich, Zurich, Switzerland
John C. Mitchell
　Stanford University, Stanford, CA, USA
Moni Naor
　Weizmann Institute of Science, Rehovot, Israel
C. Pandu Rangan
　Indian Institute of Technology, Madras, India
Bernhard Steffen
　TU Dortmund University, Dortmund, Germany
Demetri Terzopoulos
　University of California, Los Angeles, CA, USA
Doug Tygar
　University of California, Berkeley, CA, USA
Gerhard Weikum
　Max Planck Institute for Informatics, Saarbrücken, Germany

More information about this series at http://www.springer.com/series/7412

Guorong Wu · Paul Laurienti
Leonardo Bonilha · Brent C. Munsell (Eds.)

Connectomics in NeuroImaging

First International Workshop, CNI 2017
Held in Conjunction with MICCAI 2017
Quebec City, QC, Canada, September 14, 2017
Proceedings

 Springer

Editors
Guorong Wu
University of North Carolina
  at Chapel Hill
Chapel Hill, NC
USA

Paul Laurienti
Wake Forest School of Medicine
Winston-Salem, NC
USA

Leonardo Bonilha
Medical University of South Carolina
Charleston, SC
USA

Brent C. Munsell
College of Charleston
Charleston, SC
USA

ISSN 0302-9743   ISSN 1611-3349   (electronic)
Lecture Notes in Computer Science
ISBN 978-3-319-67158-1   ISBN 978-3-319-67159-8   (eBook)
DOI 10.1007/978-3-319-67159-8

Library of Congress Control Number: 2017952863

LNCS Sublibrary: SL6 – Image Processing, Computer Vision, Pattern Recognition, and Graphics

© Springer International Publishing AG 2017
This work is subject to copyright. All rights are reserved by the Publisher, whether the whole or part of the material is concerned, specifically the rights of translation, reprinting, reuse of illustrations, recitation, broadcasting, reproduction on microfilms or in any other physical way, and transmission or information storage and retrieval, electronic adaptation, computer software, or by similar or dissimilar methodology now known or hereafter developed.
The use of general descriptive names, registered names, trademarks, service marks, etc. in this publication does not imply, even in the absence of a specific statement, that such names are exempt from the relevant protective laws and regulations and therefore free for general use.
The publisher, the authors and the editors are safe to assume that the advice and information in this book are believed to be true and accurate at the date of publication. Neither the publisher nor the authors or the editors give a warranty, express or implied, with respect to the material contained herein or for any errors or omissions that may have been made. The publisher remains neutral with regard to jurisdictional claims in published maps and institutional affiliations.

Printed on acid-free paper

This Springer imprint is published by Springer Nature
The registered company is Springer International Publishing AG
The registered company address is: Gewerbestrasse 11, 6330 Cham, Switzerland

Preface

The First International Workshop on Connectomics in NeuroImaging (CNI 2017) was held in Quebec City, Canada, on September 14th, 2017, in conjunction with the 20th International Conference on Medical Image Computing and Computer Assisted Intervention (MICCAI).

Connectomics is the study of whole brain maps of connectivity, commonly referred to as the brain connectome, which focuses on quantifying, visualizing, and understanding brain network organization, including its applications in neuroimaging. The primary academic objective is to bring together computational researchers (computer scientists, data scientists, and computation neuroscientists) to discuss new advancements in network construction, analysis, and visualization techniques in connectomics and their use in clinical diagnosis and group comparison studies. The secondary academic objective to attract neuroscientists and clinicians to show recent methodological advancements in connectomics, and how they are successfully applied in various neuroimaging applications. CNI 2017 was held as a single-track workshop, which included: four keynote speakers (Bharat Biswal, Chris Rorden, Boris Bernhardt, and Moo Chung), oral paper presentations, poster sessions, and software demonstrations.

The quality of submissions to our workshop was very high. Authors were asked to submit 8 pages in LNCS format for review. A total of 26 papers were submitted to the workshop in response to the call for papers. Each of the 26 papers underwent a rigorous double-blind peer-review process, with each paper being reviewed by at least two (typically three) reviewers from the Program Committee, which was composed of 31 well-known experts in the field of connectomics. Based on the reviewing scores and critiques, the best 19 papers were accepted for presentation at the workshop, and chosen to be included in this Springer LNCS volume. The large variety of connectomics techniques, applied in neuroimaging applications, were well represented at the CNI 2017 workshop.

We are grateful to the Program Committee for reviewing the submitted papers and giving constructive comments and critiques, to the authors for submitting high-quality papers, to the presenters for excellent presentations, and to all the CNI 2017 attendees who came to Quebec City from all around the world.

September 2017

Guorong Wu
Paul Laurienti
Leonardo Bonilha
Brent Munsell

Organization

Program Committee

Pierre Besson	Aix-Marseille Université, France
Sylvain Bouix	Harvard Medical School, USA
Dante Chialvo	Universidad Nacional de San Martin, Argentine
Ai Wern Chung	Harvard Medical School, USA
Jessica Cohen	University of North Carolina, USA
Eran Dayan	University of North Carolina, USA
Simon Davis	Duke University, USA
Maxime Descôteaux	Université de Sherbrooke, Canada
Yong Fan	University of Pennsylvania, USA
Wei Gao	Cedars-Sinai Hospital, USA
Ghassan Hamarneh	Simon Fraser University, Canada
Yong He	Beijing Normal University, China
Daniel Kaufer	University of North Carolina, USA
Renaud Lopes	University of Lille Nord de France, France
Barbara Marebwa	Medical University of South Carolina, USA
Emilie Mckinnon	Medical University of South Carolina, USA
Vinod Menon	Stanford University, USA
Iman Mohammad-Rezazadeh	University of California Los Angeles, USA
Lauren O'Donnell	Harvard Medical School, USA
Ziwen Peng	Southern China Normal University, China
Luiz Pessoa	University of Maryland, USA
Islem Rekik	University of Dundee, UK
Mert Sabuncu	Cornell University, USA
Dustin Scheinost	Yale University, USA
Markus Schirmer	Harvard Medical School, USA
Li Shen	Indiana University, USA
Martha Shenton	Harvard Medical School, USA
Martin Styner	University of North Carolina, USA
Heung-ll Suk	Korea University, South Korea
Yihong Yang	NIH/NIDA, USA
Hungtu Zhu	University of Texas MD Anderson Cancer Center, USA

Contents

Connectome of Autistic Brains, Global Versus Local Characterization   1
  Saida S. Mohamed, Nancy Duong Nguyen, Eiko Yoneki,
  and Alessandro Crimi

Constructing Multi-frequency High-Order Functional Connectivity
Network for Diagnosis of Mild Cognitive Impairment   9
  Yu Zhang, Han Zhang, Xiaobo Chen, and Dinggang Shen

Consciousness Level and Recovery Outcome Prediction
Using High-Order Brain Functional Connectivity Network 17
  Xiuyi Jia, Han Zhang, Ehsan Adeli, and Dinggang Shen

Discriminative Log-Euclidean Kernels for Learning on Brain Networks 25
  Jonathan Young, Du Lei, and Andrea Mechelli

Interactive Computation and Visualization of Structural Connectomes
in Real-Time 35
  Maxime Chamberland, William Gray, Maxime Descoteaux,
  and Derek K. Jones

Pairing-based Ensemble Classifier Learning using Convolutional
Brain Multiplexes and Multi-view Brain Networks
for Early Dementia Diagnosis 42
  Anna Lisowska, Islem Rekik, and The Alzheimers Disease
  Neuroimaging Initiative

High-order Connectomic Manifold Learning for Autistic
Brain State Identification 51
  Mayssa Soussia and Islem Rekik

A Unified Bayesian Approach to Extract Network-Based Functional
Differences from a Heterogeneous Patient Cohort 60
  Archana Venkataraman, Nicholas Wymbs, Mary Beth Nebel,
  and Stewart Mostofsky

FCNet: A Convolutional Neural Network for Calculating Functional
Connectivity from Functional MRI 70
  Atif Riaz, Muhammad Asad, S.M. Masudur Rahman Al-Arif,
  Eduardo Alonso, Danai Dima, Philip Corr, and Greg Slabaugh

Identifying Subnetwork Fingerprints in Structural Connectomes:
A Data-Driven Approach   79
    Brent C. Munsell, Eric Hofesmann, John Delgaizo, Martin Styner,
    and Leonardo Bonilha

A Simple and Efficient Cylinder Imposter Approach to
Visualize DTI Fiber Tracts   89
    Lucas L. Nesi, Chris Rorden, and Brent C. Munsell

Revisiting Abnormalities in Brain Network Architecture Underlying
Autism Using Topology-Inspired Statistical Inference   98
    Sourabh Palande, Vipin Jose, Brandon Zielinski, Jeffrey Anderson,
    P. Thomas Fletcher, and Bei Wang

"Evaluating Acquisition Time of rfMRI in the Human Connectome
Project for Early Psychosis. How Much Is Enough?"   108
    Sylvain Bouix, Sophia Swago, John D. West, Ofer Pasternak,
    Alan Breier, and Martha E. Shenton

Early Brain Functional Segregation and Integration Predict Later Cognitive
Performance   116
    Han Zhang, Weiyan Yin, Weili Lin, and Dinggang Shen

Measuring Brain Connectivity via Shape Analysis of fMRI Time Courses
and Spectra   125
    David S. Lee, Amber M. Leaver, Katherine L. Narr, Roger P. Woods,
    and Shantanu H. Joshi

Topological Network Analysis of Electroencephalographic Power Maps   134
    Yuan Wang, Moo K. Chung, Daniela Dentico, Antoine Lutz,
    and Richard J. Davidson

Region-Wise Stochastic Pattern Modeling for Autism Spectrum Disorder
Identification and Temporal Dynamics Analysis   143
    Eunji Jun and Heung-Il Suk

A Whole-Brain Reconstruction Approach for FOD Modeling
from Multi-Shell Diffusion MRI   152
    Wei Sun, Junling Li, and Yonggang Shi

Topological Distances Between Brain Networks   161
    Moo K. Chung, Hyekyoung Lee, Victor Solo, Richard J. Davidson,
    and Seth D. Pollak

Author Index   171

Connectome of Autistic Brains, Global Versus Local Characterization

Saida S. Mohamed[1,5(✉)], Nancy Duong Nguyen[1,3], Eiko Yoneki[4],
and Alessandro Crimi[2]

[1] African Institute for Mathematical Sciences of Tanzania, Bagamoyo, Tanzania
[2] African Institute for Mathematical Sciences of Ghana, Biriwa, Ghana
[3] School of Mathematics and Statistics, University College Dublin, Dublin, Ireland
[4] Computer Laboratory, University of Cambridge, Cambridge, UK
[5] Faculty of Science, Cairo University, Giza, Egypt
saida@sci.cu.edu.eg

Abstract. The underlying neural mechanisms of autism spectrum disorders (ASD) remains unclear. Most of the previous studies based on connectomics to discriminate ASD from typically developing (TD) subjects focused either on global graph metrics or specific discriminant connections. In this paper we investigate whether there is a correlation between local and global features, and whether the characterization that discriminates ASD from TD subjects is primarily given by widespread network differences, or the difference lies in specific local connections which are just captured by global metrics. Namely, whether miswiring of brain connections related to ASD is localized or diffuse. The presented results suggest that the widespread hypothesis is more likely.

Keywords: ASD · Connectome · Tractography · Autism · Graph metrics

1 Introduction

A connectome is a mathematical representation of the brain as a network comprising a set of nodes and edges that relate them [17]. Nodes represent distinct homogeneous brain regions generally defined by a brain atlas. Edges represent connectivity, either functional given by co-activation in time of functional signal, or structural given by the fibers physically connecting the areas. Some brain pathologies investigated by using connectomes have been considered either by their effect in specific local connections or by their impact to the global brain network. For instance, with Alzheimer's disease there is an overall disruption of structural and functional connectivity [13]. Schizophrenia is considered the "disconnection" disease with several miswirings between brain areas [20]. Stroke and gliomas are mostly focal lesions and many studies have shown disruptions in structural and functional connectivity related to the focal damage, though subsequent changes on the global organization might be present [9].

© Springer International Publishing AG 2017
G. Wu et al. (Eds.): CNI 2017, LNCS 10511, pp. 1–8, 2017.
DOI: 10.1007/978-3-319-67159-8_1

1.1   Connectomes and Autism Spectrum Disorder

Autism spectrum disorder (ASD) is a set of neuro-developmental disorders characterized by impaired social interaction and repetitive behaviors [1]. The underlying neural mechanism of ASD remains unclear. Magnetic resonance imaging-based characterization of ASD has been explored as a complement to the current behavior-based diagnoses [19]. Several studies have proposed biomarkers for discrimination of ASD subjects. Rudie et al. investigated global metrics obtained from functional and structural connectomes [16]. The same metrics have also been used in a support vector machine (SVM) framework to characterize global changes in the connectome of ASD subjects [7]. A *rich-club* refers to a close group of nodes with relatively high degree. Ray et al. used an overall rich-club score for the connectome to discriminate ASD, attention deficit/hyperactive disorder, and typically developing (TD) subjects [14]. At the local level, ASD has been investigated looking for few connections which can be used to discriminate ASD from TD subjects. Promising results have been found using functional connectomes [19], structural connectomes [12], and effective connectivity graphs [3,10]. Lastly, local areas have been studied by using the same global graph metrics used in the aforementioned works but applied to specific local network regions [8], and 10 areas were found statistically different among ASD and TD subject groups. Nevertheless, given the high number of needed connections to discriminate between the two groups in a case-control setting, the question remains whether the most representative biomarkers, which allow the discrimination, are specific local connection differences or ASD is a diffuse global connectome disconnection pathology such as schizophrenia. In this paper we want to investigate whether there is a correlation between these two aspects, and whether one is more predominant than the other. To do so, we compute the most common global metrics and verify if any of them is useful in discriminating ASD from TD subjects. We then seek for the local connections which are different across those two groups and whether there is a correlation between the two types of characterization. In the end, we draw some conclusion considering all these results. The rationale is that if there is statistically significant global metric that discriminate the two groups, there might be a correlation either with single specific local features or with the ensemble of local features.

1.2   Global Metrics

Global metrics are important tools to analyse the network because they allow us to represent with few scalar values the topology and efficiency of a network. Those might represent the segregation, integration, centrality, and resilience of a network. To be in line with previous works on ASD [7,16], we focus on network segregation and integration, using only features which are statistically representative for our dataset.

– Segregation refers to the process of grouping communities such that members of the same community are more densely connected than members of different communities. This is similar to the concept of clustering and community detection [4].

– Integration refers to the network's ability to propagate information and the efficiency of global communication [4].

In our experiments, we tested several metrics of integration and segregation for weighted graphs in discriminating the two groups with a t-test, and we then retained those which are statistically significant (p-value < 0.5). Those are one metric of segregation (Louvain modularity) and one of integration (characteristic path length) both in their weighted version.

The *Louvain modularity* method is a community detection method that partitions the network using a greedy algorithm that optimizes the modularity [15]. The optimization is performed in two steps. First, the method groups individual nodes into "small" communities by optimizing modularity locally. Second, it builds a new network whose nodes are the newly formed communities. These steps are iterated until a maximum of modularity is attained and a hierarchy of communities is produced. For weighted graphs, modularity is defined as

$$Q = \frac{1}{2m} \sum_{ij} \left[A_{ij} - \frac{k_i k_j}{2m} \right] \delta(c_i, c_j), \tag{1}$$

where A_{ij} is the weight of the edge connecting between nodes i and j from the adjacency matrix $\mathbf{A}$, k_i and k_j are the sums of weights of the edges connected to node i and j respectively, $m = 1/(2A_{ij})$, c_i and c_j are the communities of nodes i and j, and δ is a simple delta function.

Weighted characteristic path length measures the integrity of the network and the ease of information flow within the network. The distance d_{ij} is the shortest path between node i and j. It is quantified by the weighted count of edges in this shortest path [15]. The characteristic path length is the average of all the distances between every pair in the network defined as

$$L^W = \frac{1}{n(n-1)} \sum_{i,j \in n, i \neq j} d_{ij}^W, \tag{2}$$

where n is the number of nodes.

1.3 Local Connectivity Differences

We define specific connections which can discriminate between 2 groups of networks as *local connectivity difference*. Local connectivity difference can be found in several ways, as false discovery rate [20], by analyzing the SVM weights trained to discriminate between ASD and TD [6], or by using network based statistics (NBS) [20]. NBS is a nonparametric statistical test used to identify connections within connectivity matrices which are statistically significant different between two distinct populations [20]. In practice, the NBS checks the family-wise error rate, where the null hypothesis is tested independently at each of the edges. This is achieved performing a two-sample t-test at each edge independently using the values of connectivity. The tests are repeated k times, each time randomly permuting members of the two populations.

2   Methods

Our evaluation is carried out with the following steps:

1. For all connectomes of both ASD and TD subjects the aforementioned global metrics are computed.
2. Those metrics are used as features for an SVM classification task.
3. NBS is performed to identify discriminant local connection.
4. Local and global features are then compared.

Beyond the SVM classification, to extract more meaning from the features as in [16] a univariate t-test is performed on the single features assessing their statistical significance independently from the other features. To compare global and local features, univariate and multivariate regressions are performed between the statistically significant global metrics and the local connections.

3   Data and Experimental Settings

The experiments have been performed on the San Diego State University cohort of the ABIDE-II dataset [5]. This cohort was chosen as it was the one with diffusion tensor imaging (DTI) volumes at sufficient resolution to allow acceptable quality tractography. One sample was discarded as it produced too noisy tractography with the used algorithm. The final dataset included 30 ASD and 24 TD subjects matched for age, gender, handedness, and nonverbal intelligence quotient. For each subject, DTI and T1 have been acquired and co-registered. Imaging data were acquired on a GE (Milwaukee, WI) 3T MR750 scanner. T1 data were acquired with repetition time (TR) = 8.108 ms, echo time (TE) = 3.172 ms, flip angle = 8°, 172 slices, 1 mm³ resolution. DTI volumes were obtained with an echo-planar pulse sequence with full head coverage and encoded for 61 noncollinear diffusion directions with TR = 8,500 ms, TE = 84.9 ms, flip angle = 90°, FOV = 240 mm, 128 × 128 matrix, 1.88 × 1.88 × 2 mm³ resolution.

3.1   Pre-processing and Connectome Construction

DTI volumes have been pre-processed with eddy current correction and skull stripping. Linear registration has been applied between the automated anatomic labeling (AAL) atlas [18] and the T1 reference volume by using linear registration with 12 degrees of freedom. Tractographies for all subjects have been generated processing DTI data with a deterministic Euler approach stemming from 2,000,000 seed-points and stopping when the fractional anisotropy was smaller than <0.1. Additionally, all the structural connections with fiber lengths <30 mm were also excluded. To construct the connectome, the graph nodes have been determined using the 90 regions in the AAL atlas. The edges have been weighted with the number of tracts connecting two regions.

3.2   Experimental Settings

All features are computed by using the Brain Connectivity Toolbox[1], while the SVM implementation of Scikit-learn[2] was used. Due to the stochastic aspects of the Louvain modularity, experiments were repeated 100 times, averaging the results. NBS was used with $k = 1000$ permutations thresholding the p-value at $\alpha = 0.01$. All results are computed in a leave-one-out cross-validation manner.

4   Results and Discussions

Classification performance by using the global metrics jointly and an SVM classifier can be summarized by the receiver operating characteristic curve (ROC) shown in Fig. 1(a). It can be seen that the mean of 100 runs is significant with a mean area under the curve (AUC) of 0.77 in agreement with similar previous results on another dataset [7]. Performing the SVM classification using only one feature at time, the AUC was 0.70 (mean) for the Louvain modularity, and 0.74 for the characteristic path length. It is worthwhile to mention that the Louvain modularity was producing sometimes relatively high and sometimes relatively low AUC. Table 1 shows the mean and standard deviation for each feature for ASD and TD brain matrices, and the resulting p-values indicate the features where the two classes differ. Those results are in agreement with Rudie et al. [16]. The difference in structural modularity and characteristic path length among ASD and TD subjects can reflect a subtle randomization of the network connectivity as proposed in [15]. Despite using the Louvain modularity and characteristic path length as a representation of segregation and integration can be reductive, giving their significance with the used dataset we used those features jointly with local differences looking for correlations. NBS detected 10 symmetric discriminant structural connections depicted in Fig. 1(b–d), are similar to those obtained with the same dataset and using SVM in [3], and on another dataset using also NBS [8]. However, those are slightly different from the functional connections detected in [19].

Table 1. Mean value and standard deviation of the global metrics for both the ASD and TD population. The last column gives the one-tail p-values comparing the two.

Feature	ASD		TD		p-value
	Mean	Std	Mean	Std	
Louvain modularity	0.542	±0.020	0.532	±0.021	0.047
Weighted characteristic path length	0.0163	±0.0016	0.0155	±0.0010	0.049

It is worthwhile to mention that the detected connections are obtained with a p-value threshold of 0.01 and with a p-value threshold of 0.05 there are approximately 3 times more connections. Those connections are mostly located at the

---

[1] http://brain-connectivity-toolbox.net.

[2] http://scikit-learn.org/.

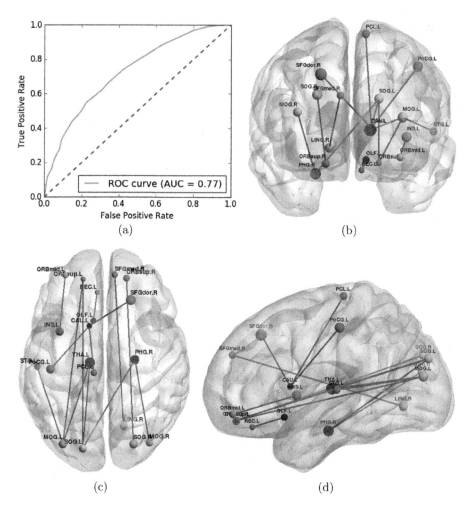

Fig. 1. (a) Mean ROC and AUC for the classification task with both features obtained averaging 100 times the ROC with different Louvain modularity computed. (b) axial, (c) coronal and (d) sagittal view of the statistical different connections between the ASD and TD subjects. The used abbreviations are MOG = middle occipital gyrus, SOG = superior occipital gyrus, PoCG = posterior cingulate gyrus, THA = thalamus, SFG = superior frontal gyrus, INS = insula, PCL = Paracentral lobule, CAU = caudate, PHG = para hippocampal gyrus, L = left, and R= right.

occipital gyrus on both sides going to the orbitofrontal cortex, thalamus and caudate left, para-hippocampal gyrus, in agreement with former studies [8,11]. The accuracy of the SVM classification using the 2 global features was 65% while using the 10 discriminant connection was 61%. This is in line with previous studies [7], which are also suggesting that accuracy can be increased if functional rather than structural connectivity is used [19].

The univariate correlation between the mean Louvain modularity or weighted characteristic path length and single connections detected by the NBS was not so strong. In fact, all the computed r^2 score were between 0.01 and 0.2, when 0 is no correlation and 1 is perfect correlation. Instead, performing a multivariate regression between the Louvain modularity or path length and all connectivity values jointly gave respectively a mean $r^2 = 0.49$ and $r^2 = 0.48$, suggesting a stronger meaning in using all connections jointly. Moreover, given also the high number of different connections between the two groups (10 and 31 for the $\alpha = 0.01$ and $\alpha = 0.05$ p-value threshold respectively), and their location spread across the brain, we conclude that ASD is more characterized by spread miswiring similar to schizophrenia rather than few representative disconnections. Therefore, we agree with previous hypothesis that a global disruption of connectivity is the basis of ASD, and that changes during development compensate for the disruption [2], although the experiments should be repeated with larger sample sizes.

5 Conclusion

In this paper we have shown that the structural connectome of ASD and TD subjects can be classified by either using the Louvain modularity and characteristic path length or a set of structural connections, giving similar accuracy. Lastly, given still the high number of connections and their heterogeneous location within the brain of structural connection, ASD could be considered as a widespread miswiring of the brain. Future work comprises the use of functional connectivity in the analysis, the inclusion of other global metrics beyond segregation and integration, and investigating different settings for classification.

References

1. Association, A.P., et al.: Diagnostic and Statistical Manual of Mental Disorders (DSM-5®). American Psychiatric Pub (2013)
2. Belmonte, M., Allen, G., Beckel-Mitchener, A., Boulanger, L., Carper, R., Webb, S.: Autism and abnormal development of brain connectivity. J. Neurosci. **24**, 9228–9231 (2004)
3. Crimi, A., Dodero, L., Murino, V., Sona, D.: Case-control discrimination through effective brain connectivity. In: IEEE ISBI (2017)
4. Deco, G., Tononi, G., Boly, M., Kringelbach, M.L.: Rethinking segregation and integration: contributions of whole-brain modelling. Nat. Rev. Neurosci. **16**(7), 430–439 (2015)
5. Di Martino, A., O'Connor, D., Chen, B., Alaerts, K., Anderson, J.S., Assaf, M., Balsters, J.H., Baxter, L., Beggiato, A., Bernaerts, S., et al.: Enhancing studies of the connectome in autism using the autism brain imaging data exchange II. Sci. Data **4**, 170010 (2017)
6. Gaonkar, B., Davatzikos, C.: Analytic estimation of statistical significance maps for support vector machine based multi-variate image analysis and classification. Neuroimage **78**, 270–283 (2013)

7. Goch, C., et al.: Global changes in the connectome in autism spectrum disorders. In: Schultz, T., Nedjati-Gilani, G., Venkataraman, A., O'Donnell, L., Panagiotaki, E. (eds.) Computational Diffusion MRI and Brain Connectivity. MV, pp. 239–247. Springer, Cham (2014). doi:10.1007/978-3-319-02475-2_22

8. Li, H., Xue, Z., Ellmore, T.M., Frye, R.E., Wong, S.T.: Network-based analysis reveals stronger local diffusion-based connectivity and different correlations with oral language skills in brains of children with high functioning autism spectrum disorders. Hum. Brain Mapp. **35**(2), 396–413 (2014)

9. Lim, J.S., Kang, D.W.: Stroke connectome and its implications for cognitive and behavioral sequela of stroke. J. Stroke **17**(3), 256–267 (2015)

10. Munsell, B.C., Wu, G., Gao, Y., Desisto, N., Styner, M.: Identifying relationships in functional and structural connectome data using a hypergraph learning method. In: Ourselin, S., Joskowicz, L., Sabuncu, M.R., Unal, G., Wells, W. (eds.) MICCAI 2016. LNCS, vol. 9901, pp. 9–17. Springer, Cham (2016). doi:10.1007/978-3-319-46723-8_2

11. Nair, A., Treiber, J.M., Shukla, D.K., Shih, P., Müller, R.A.: Impaired thalamocortical connectivity in autism spectrum disorder: a study of functional and anatomical connectivity. Brain **136**(6), 1942–1955 (2013)

12. Owen, J.P., Li, Y.O., Ziv, E., Strominger, Z., Gold, J., Bukhpun, P., Wakahiro, M., Friedman, E.J., Sherr, E.H., Mukherjee, P.: The structural connectome of the human brain in agenesis of the corpus callosum. Neuroimage **70**, 340–355 (2013)

13. Pievani, M., de Haan, W., Wu, T., Seeley, W., Frisoni, G.: Functional network disruption in the degenerative dementias. Lancet Neurol. **10**, 829–843 (2011)

14. Ray, S., Miller, M., Karalunas, S., Robertson, C., Grayson, D.S., Cary, R.P., Hawkey, E., Painter, J.G., Kriz, D., Fombonne, E., et al.: Structural and functional connectivity of the human brain in autism spectrum disorders and attention-deficit/hyperactivity disorder: a rich club-organization study. Hum. Brain Mapp. **35**(12), 6032–6048 (2014)

15. Rubinov, M., Sporns, O.: Complex network measures of brain connectivity: uses and interpretations. Neuroimage **52**(3), 1059–1069 (2010)

16. Rudie, J.D., Brown, J., Beck-Pancer, D., Hernandez, L., Dennis, E., Thompson, P., Bookheimer, S., Dapretto, M.: Altered functional and structural brain network organization in autism. NeuroImage Clin. **2**, 79–94 (2013)

17. Sporns, O.: Network attributes for segregation and integration in the human brain. Curr. Opin. Neurobiol. **23**(2), 162–171 (2013)

18. Tzourio-Mazoyer, N., Landeau, B., Papathanassiou, D., Crivello, F., Etard, O., Delcroix, N., Mazoyer, B., Joliot, M.: Automated anatomical labeling of activations in SPM using a macroscopic anatomical parcellation of the MNI MRI single-subject brain. Neuroimage **15**(1), 273–289 (2002)

19. Yahata, N., Morimoto, J., Hashimoto, R., Lisi, G., Shibata, K., Kawakubo, Y., Kuwabara, H., Kuroda, M., Yamada, T., Megumi, F., et al.: A small number of abnormal brain connections predicts adult autism spectrum disorder. Nat. Commun. **7** (2016)

20. Zalesky, A., Fornito, A., Bullmore, E.T.: Network-based statistic: identifying differences in brain networks. Neuroimage **53**(4), 1197–1207 (2010)

Constructing Multi-frequency High-Order Functional Connectivity Network for Diagnosis of Mild Cognitive Impairment

Yu Zhang, Han Zhang, Xiaobo Chen, and Dinggang Shen[✉]

Department of Radiology and BRIC, University of North Carolina at Chapel Hill,
Chapel Hill, USA
dgshen@med.unc.edu

Abstract. Human brain functional connectivity (FC) networks, estimated based on resting-state functional magnetic resonance imaging (rs-fMRI), has become a promising tool for imaging-based brain disease diagnosis. Conventional low-order FC network (LON) usually characterizes pairwise temporal correlation of rs-fMRI signals between any pair of brain regions. Meanwhile, high-order FC network (HON) has provided an alternative brain network modeling strategy, characterizing more complex interactions among low-order FC sub-networks that involve multiple brain regions. However, both LON and HON are usually constructed within a fixed and relatively wide frequency band, which may fail in capturing (sensitive) frequency-specific FC changes caused by pathological attacks. To address this issue, we propose a novel "multi-frequency HON construction" method. Specifically, we construct *not only* multiple frequency-specific HONs (*intra-spectrum* HONs), *but also* a series of cross-frequency inter-action-based HONs (*inter-spectrum* HONs) based on the low-order FC sub-networks constructed at different frequency bands. Both types of these HONs, together with the frequency-specific LONs, are used for the complex network analysis-based feature extraction, followed by sparse regression-based feature selection and the classification between mild cognitive impairment (MCI) patients and normal aging subjects using a support vector machine. Compared with the previous methods, our proposed method achieves the best diagnosis accuracy in early diagnosis of Alzheimer's disease.

1 Introduction

As an irreversible, severe degenerative neurological disease, Alzheimer's disease (AD) is notorious for progressive perceptive and cognitive deficits. Mild cognitive impairment (MCI) is known as an intermediate stage between normal aging and AD. Although some individuals with MCI remain stable over time, more than half of MCI subjects progress to dementia within ~5 years, at a ratio of about 10–15% per year [1]. Such a high conversion rate could possibly be reduced if receiving proper treatments in this "early AD" stage. Thus, early detection of MCI is significantly important and clinically valuable for delaying AD progression. However, accurate MCI diagnosis based on brain

© Springer International Publishing AG 2017
G. Wu et al. (Eds.): CNI 2017, LNCS 10511, pp. 9–16, 2017.
DOI: 10.1007/978-3-319-67159-8_2

imaging is still challenging, since brain anatomical and functional changes at this stage are considerably subtle [2, 3].

Resting-state functional magnetic resonance imaging (rs-fMRI), which measures the blood oxygenation level-dependent (BOLD) signals as a neurophysiological index of neural activity, has been successfully applied to identify functional pathological biomarkers for MCI diagnosis [4]. Functional connectivity (FC), defined as the temporal correlation of BOLD signals between any pair of brain regions, has been widely applied to explore brain intrinsic functional architectures, with which a whole-brain FC network can be constructed; such connectomics information has contributed considerably to brain disease diagnosis [5, 6]. While promising, the previous mostly-adopted FC network is a typical *low-order* network (LON), since it usually characterizes the pairwise relationship between brain regions by using the temporal synchronization of BOLD signals. As a result, this type of network can hardly reveal the potentially complex relationship and high-level interactions among multiple brain regions, which may be more sensitive to the subtle MCI-related changes.

One of recent promising technique advances on brain network modeling is called *high-order* FC network (HON), which quantifies high-level inter-regional interactions by using topographical resemblance information between low-order sub-networks [7, 8]. However, both LON and HON are constructed at a fixed and relatively wide frequency band, which may be insensitive and insufficient to capture the frequency-specific changes caused by early pathological attacks. Actually, it has been suggested that the neuronal oscillations at distinct frequency bands have different biophysiological meanings and may contribute differently to FC [9]. Thus, the frequency-specific FC as well as the cross-frequency interaction analysis have opened up a new effective way for exploring basic neuroscience problems on high-level cognitive functions [10], and could be used for revealing subtle pathological variations in a scenario where the traditional LON is less effective to detect [11].

In this study, we propose a novel brain connectomics-based disease diagnosis framework based on frequency-specific HONs from rs-fMRI. Specifically, we construct multiple *frequency-specific* HONs and multiple *cross-frequency interaction-based* HONs (based on the LONs calculated at each frequency band). The *frequency-specific* HONs (namely, *intra-spectrum* HONs) are constructed based on the topographical similarity between low-order sub-networks for each sub-frequency band, which characterize the *intra-spectrum* high-level interactions. On the other hand, the *cross-frequency interaction-based* HONs (namely, *inter-spectrum* HONs) are constructed by quantifying cross-frequency-band topographical similarity between low-order sub-networks derived from different frequency bands. Since different frequency bands carry different neurobiological functions, such *inter-spectrum* HONs are able to measure high-level modulations among brain functional systems. To evaluate the effectiveness of these new network modeling metrics, we use both types of these HONs, along with frequency-specific LONs, for classification between MCI patients and normal controls (NC). We conduct extensive comparisons between our framework and other state-of-the-art methods in MCI diagnosis. The results show our framework can significantly outperform comparison methods.

2   Methods

Our proposed framework consists of the following 6 steps. (1) For each subject, the regional mean time series of BOLD signals in each ROI are decomposed into multiple frequency sub-bands. (2) Within each sub-band, one LON is constructed by calculating Pearson's correlation between each pair of the frequency-specific regional mean time series. (3) An *intra-spectrum* HON is thus estimated based on the topographical similarity between each pair of the low-order FC sub-networks constructed from the same frequency sub-band. (4) An *inter-spectrum* HON is further estimated based on the topographical similarity between each pair of the low-order FC sub-networks that are constructed from two different frequency sub-bands. (5) From each of the constructed FC networks, the complex network property-related features are extracted using weighted clustering coefficients for each "node" [12]; among them, the discriminative features are selected using sparse regression-based feature selection. (6) Support vector machine (SVM) with a linear kernel is trained on the selected features for MCI classification. Figure 1 illustrates the main flowchart of our proposed multi-frequency HONs construction approach.

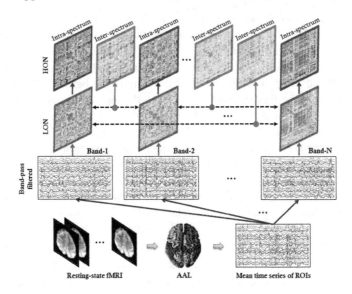

Fig. 1. Illustration on how to construct *intra-spectrum* low-order and high-order FC networks, as well as *inter-spectrum* high-order FC networks.

2.1   Multi-frequency High-Order FC Networks

Suppose that $\mathbf{X} \in \mathbb{R}^{P \times R}$ denotes regional mean time series with P time points from a total of R regions-of-interest (ROIs), where each mean time series has been *band-pass* filtered at a relatively wide frequency band. A full-spectrum FC is derived by computing the Pearson's correlation C_{ij} between the mean time series of the i-th and the j-th ROIs.

By estimating the FC between each possible pair of ROIs, a FC network can be constructed as a symmetric matrix $\mathbf{C} = [C_{ij}] \in \mathbb{R}^{R \times R}$. Without loss of generality, we assume each column of $\mathbf{X}$ has been de-meaned and also variance-normalized by dividing by the standard deviation. The FC network can thus be equivalently computed by $\mathbf{C} = \mathbf{X}^T \mathbf{X}$. This defines one LON by simply calculating temporal correlation between mean time series from any pair of brain ROIs. In addition, this LON is a full-spectrum FC network (i.e., 0.015–0.15 Hz in this study), which could be incapable to capture the subtle pathological changes particularly at specific frequency spectrum.

In this study, we address the above issue by the construction of multi-frequency HONs. By fast Fourier transformation, the mean time series $\mathbf{X}$ can be decomposed into frequency-band-specific time series $\mathbf{X}^k \in \mathbb{R}^{P \times R}$ ($k = 1, 2, 3$ and 4) at four different sub-bands: $SB_1 = 0.015–0.0488$ Hz, $SB_2 = 0.0488–0.0825$ Hz, $SB_3 = 0.0825–0.1163$ Hz, $SB_4 = 0.1163–0.15$ Hz (through equal separation of 0.015–0.15 Hz). Within the k-th sub-band, one LON can be constructed as $\mathbf{C}^k = (\mathbf{X}^k)^T \mathbf{X}^k$, and thus totally four frequency-specific LONs can be obtained. Different from full-spectrum analysis, these frequency-specific LONs (constructed at their respective sub-frequency bands) are able to reveal those frequency-specific pathological variations.

Alternatively, each frequency-specific LON $\mathbf{C}^k$ can also be rewritten as $\mathbf{C}^k = [\mathbf{c}_1^k, \mathbf{c}_1^k, \ldots, \mathbf{c}_R^k] \in \mathbb{R}^{R \times R}$, where the i-th column $\mathbf{c}_i^k$ (or the i-th row due to the symmetry of $\mathbf{C}^k$) delineates the connectivity pattern between the i-th ROI and all other ROIs and can be regarded as a low-order "sub-network". Thus, a high-order FC can further be defined as pair-wise topographical similarity between low-order sub-networks. Similar to the frequency-specific LONs, the *intra-spectrum* HON at the k-th sub-band can be constructed by calculating the high-order FC between every pair of low-order sub-networks as $\mathbf{H}^k = (\mathbf{C}^k)^T \mathbf{C}^k$. The difference between the *frequency-specific* LONs and the *intra-spectrum* HON is that the latter characterizes high-level interactions among brain regions in each frequency sub-band, and totally there are four *intra-spectrum* HONs.

To comprehensively explore high-level FC, we further construct an *inter-spectrum* HON which is calculated based on the correlation of low-order sub-networks defined at two different sub-bands. Specifically, the inter-spectrum high-order FC between two different frequency sub-bands, i.e., SB_k and SB_l, is estimated by computing Pearson's correlation between $\mathbf{c}_i^k$ and $\mathbf{c}_j^l$ ($k, l = 1, \ldots, 4, k \neq l; i, j = 1, \ldots, R, i \neq j$). With a unified form, an *inter-spectrum* HON between two different sub-bands can be constructed by $\mathbf{H}^{kl} = (\mathbf{C}^k)^T \mathbf{C}^l$. Such *inter-spectrum* HON provides a straightforward way to characterize the high-level cross-frequency modulations among brain regions. For the four sub-frequency bands, we have totally $C_4^2 = 6$ inter-spectrum HONs.

2.2   Feature Extraction and Classification

For each subject, a total of 14 FC networks are constructed, including (1) four *frequency-specific* LONs, (2) four *intra-spectrum* HONs, and (3) six *inter-spectrum* HONs. But, the feature dimensionality will be rather high, if directly using the connectivity strengths from these 14 networks as features. An effective alternative is using complex network properties, extracted by graph theoretic analysis, as high-level features. To this end, we compute a weighted local clustering coefficient [12, 13] for each node (for reflecting the efficiency of information transferring in a local range) in each network as a feature, and then concatenate all these features to form a long feature vector with the length of $14 \times R$. Since there could be some redundant features which may affect classification, we conduct feature selection based on sparse regression [14, 15] to derive a subset of features with best discriminability. Finally, the SVM with a linear kernel is trained on the selected feature subset for MCI classification.

3   Experiments

3.1   Data

We use the Alzheimer's Disease Neuroimaging Initiative (ADNI) dataset (http:// adni.loni.usc.edu/) for validation of the proposed multi-frequency HONs in MCI classification. Totally, 59 NC subjects and 53 MCI patients (consisting of both early and late MCIs) are selected from ADNI-2 for our experiments. Subjects from both classes are age- and gender-matched, and they were all scanned using 3.0T Philips scanners. The rs-fMRI data are preprocessed using SPM8 software (http://www.fil.ion.ucl.ac.uk/ spm/software/spm8/) according to the well-accepted pipeline. Specifically, the first three volumes of each subject are discarded before preprocessing for magnetization equilibrium. Then, rigid-body registration is used to correct head motion. The rs-fMRI data are normalized to Montreal Neurological Institute (MNI) space, and further spatially smoothed by a Gaussian kernel with full-width-at-half-maximum (FWHM) of $6 \times 6 \times 6 \, mm^3$. Of note, we do not perform scrubbing to the data with large (i.e., >0.5 mm) frame-wise displacement. However, the subjects who have more than 2.5 min rs-fMRI data with large frame-wise displacement are excluded for further analysis. Head motion parameters and also the mean BOLD time series of white matter and cerebrospinal fluid are regressed out to further remove artifacts that may interfere with FC estimation. According to the Automated Anatomical Labeling (AAL) atlas, the rs-fMRI data are parcellated into 116 ROIs. Regional mean rs-fMRI time series of each ROI is *band-pass* filtered between 0.015 and 0.15 Hz.

3.2   Performance Evaluation

The leave-one-out cross-validation (LOOCV) scheme is adopted to evaluate the diagnosis performance of the proposed method. Specifically, in each fold of LOOCV procedure, an additional inner LOOCV is carried out on the training data to determine the optimal hyper-parameters for both sparse regression (used for feature selection) and SVM (used for classification). The classification performance is measured based on classification accuracy (ACC), area under ROC curve (AUC), sensitivity (SEN), and specificity (SPE). To fairly evaluate the effectiveness of our proposed framework, extensive experimental comparisons are carried out based on the following 9 methods (1) LON^F: Low-order FC networks constructed using *full-spectrum* BOLD signals; this is the most widely used method. (2) HON^F: High-order FC networks constructed based on the LON^F. (3) LON^F+HON^F: Combination of the *full-spectrum* low- and high-order FC networks. (4) LON^{IA}: Intra-spectrum low-order FC networks, which were previously used mainly for group-level analysis. (5) HON^{IA}: Intra-spectrum high-order FC networks, newly proposed by us. (6) Inter-spectrum high-order FC networks, newly proposed. (7) LON^{IA}+HON^{IA}: Combination of the intra-spectrum low- and high-order FC networks. (8) HON^{IA}+HON^{IE}: Combination of the intra-spectrum and inter-spectrum high-order FC networks. (9) LON^{IA}+HON^{IA}+HON^{IE}: Combination of intra-spectrum low- and high-order FC networks as well as inter-spectrum high-order FC networks, i.e., our full method.

Table 1 summarizes the classification performance on MCI diagnosis for all of the 9 aforementioned methods. Compared with LONs, HONs achieved better classification performance in either *full-spectrum* or *multi-spectrum* FC analysis. From another aspect, *multiple-spectrum* FC analysis outperformed *full-spectrum* FC analysis for either LONs or HONs. By exploiting the high-level interactions among brain regions across different frequency spectrums, the HON^{IE} produced the best performance among all comparison methods using a single type of the FC networks. On the other hand, integrating different types of FC networks further improved the classification results. The combination of LON^{IA}, HON^{IA} and HON^{IE} yielded the best classification performance (i.e., 83.9% in accuracy). This indicates that all the three types of FC networks, characterizing brain functional organizations from different aspects, provide complementary information to each other for MCI diagnosis.

Table 1. Performance comparison of different methods in MCI classification.

Method	ACC (%)	AUC	SEN (%)	SPE (%)
LON^F	61.6	0.648	56.6	66.1
HON^F	65.2	0.658	56.6	72.9
LON^F + HON^F	67.9	0.698	60.4	74.6
LON^{IA}	70.5	0.746	66.0	74.6
HON^{IA}	73.2	0.747	71.7	74.6
HON^{IE}	75.0	0.757	71.7	78.0
LON^{IA} + HON^{IA}	75.9	0.798	73.6	78.0
HON^{IA} + HON^{IE}	79.5	0.833	77.4	81.4
LON^{IA} + HON^{IA} + HON^{IE}	**83.9**	**0.908**	**79.3**	**88.1**

3.3   Intra-spectrum and Inter-spectrum HONs

Figure 2 presents the group-averaged HON^{IA} (0.015–0.0488 Hz), HON^{IA} (0.0488–0.0825 Hz), and HON^{IE} (across two sub-bands) for NC and MCI groups as an example. The discriminability index [16], calculated as an r^2-value for each connection in each type of the FC network, is also shown. Larger r^2-value indicates higher separability of the feature distribution patterns between two classes. From these results, we can see that the three HONs identified several different discriminative FC links, indicating that they may serve as complementary features for MCI diagnosis. This also offers an additional evidence for the highest classification performance achieved by the combination of three types of networks.

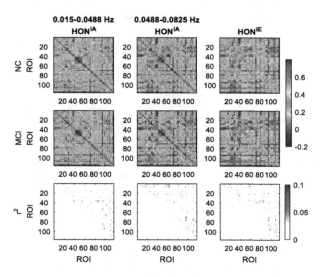

Fig. 2. Group-averaged FC networks of HON^{IA} (0.015–0.0488 Hz), HON^{IA} (0.0488–0.0825 Hz), and HON^{IE} (across two sub-bands) for NC and MCI groups, as well as the separability matrices between two groups for each type of the networks.

4   Conclusion

In this paper, we have presented a novel framework based on multi-frequency high-order FC networks for MCI diagnosis. Rather than using the full-spectrum FC, we construct both *intra-spectrum* HONs and *inter-spectrum* HONs to capture those previously ignored frequency-dependent high-order FC and cross-frequency modulation-related high-order FC. Both multi-frequency LONs and HONs are jointly used for MCI diagnosis. Experimental results show that different brain networks do provide valuable complementary information for MCI classification, and our full method achieves the best performance. This indicates the promise of the proposed brain network modeling method for brain connectomics-orientated studies.

Acknowledgements. This work is partially supported by NIH grants (EB006733, EB008374, EB009634, MH107815, AG041721, and AG042599).

References

1. Gauthier, S., Reisberg, B., Zaudig, M., Petersen, R.C., Ritchie, K., Broich, K., Belleville, S., Brodaty, H., Bennett, D., Chertkow, H., Cummings, J.L.: Mild cognitive impairment. Lancet **367**(9518), 1262–1270 (2006)

2. Zhu, X., Suk, H.I., Lee, S.W., Shen, D.: Subspace regularized sparse multitask learning for multiclass neurodegenerative disease identification. IEEE Trans. Biomed. Eng. **63**(3), 607–618 (2016)
3. Zhu, X., Suk, H.I., Wang, L., Lee, S.W., Shen, D.: A novel relational regularization feature selection method for joint regression and classification in AD diagnosis. Med. Image Anal. **38**, 205–214 (2017)
4. Allen, E.A., Damaraju, E., Plis, S.M., Erhardt, E.B., Eichele, T., Calhoun, V.D.: Tracking whole-brain connectivity dynamics in the resting state. Cereb. Cortex **24**, 663–676 (2012)
5. Chen, X., Zhang, H., Lee, S.-W., Shen, D.: Hierarchical high-order functional connectivity networks and selective feature fusion for MCI classification. Neuroinformatics 1–14 (2017)
6. Wang, J., Wang, Q., Peng, J., Nie, D., Zhao, F., Kim, M., Zhang, H., Wee, C.Y., Wang, S., Shen, D.: Multi-task diagnosis for autism spectrum disorders using multi-modality features: a multi-center study. Hum. Brain Mapp. **38**(6), 3081–3097 (2017)
7. Zhang, H., Chen, X., Shi, F., Li, G., Kim, M., Giannakopoulos, P., Haller, S., Shen, D.: Topographical information-based high-order functional connectivity and its application in abnormality detection for mild cognitive impairment. J. Alzheimers Dis. **54**(3), 1095–1112 (2016)
8. Zhang, Y., Zhang, H., Chen, X., Lee, S.-W., Shen, D.: Hybrid high-order functional connectivity networks using resting-state functional MRI for mild cognitive impairment diagnosis. Scientific Reports (2017)
9. Salvador, R., Martinez, A., Pomarol-Clotet, E., Gomar, J., Vila, F., Sarro, S., Capdevila, A., Bullmore, E.: A simple view of the brain through a frequency-specific functional connectivity measure. NeuroImage **39**(1), 279–289 (2008)
10. Tewarie, P., Hillebrand, A., van Dijk, B.W., Stam, C.J., O'Neill, G.C., Van Mieghem, P., Meier, J.M., Woolrich, M.W., Morris, P.G., Brookes, M.J.: Integrating cross-frequency and within band functional networks in resting-state MEG: a multi-layer network approach. NeuroImage **142**, 324–336 (2016)
11. Wee, C.Y., Yap, P.T., Denny, K., Browndyke, J.N., Potter, G.G., Welsh-Bohmer, K.A., Wang, L., Shen, D.: Resting-state multi-spectrum functional connectivity networks for identification of MCI patients. PLoS ONE **7**(5), e37828 (2012)
12. Rubinov, M., Sporns, O.: Complex network measures of brain connectivity: uses and interpretations. NeuroImage **52**(3), 1059–1069 (2010)
13. Chen, X., Zhang, H., Gao, Y., Wee, C.Y., Li, G., Shen, D.: High-order resting-state functional connectivity network for MCI classification. Hum. Brain Mapp. **37**(9), 3282–3296 (2016)
14. Zhang, Y., Zhou, G., Jin, J., Zhao, Q., Wang, X., Cichocki, A.: Aggregation of sparse linear discriminant analysis for event-related potential classification in brain-computer interface. Int. J. Neural Syst. **24**(1), 1450003 (2014)
15. Zhang, Y., Zhou, G., Jin, J., Zhao, Q., Wang, X., Cichocki, A.: Sparse Bayesian classification of EEG for brain-computer interface. IEEE Trans. Neural Netw. Learn. Syst. **27**(11), 2256–2267 (2016)
16. Zhang, Y., Wang, Y., Jin, J., Wang, X.: Sparse Bayesian learning for obtaining sparsity of EEG frequency bands based feature vectors in motor imagery classification. Int. J. Neural Syst. **27**(2), 1650032 (2017)

Consciousness Level and Recovery Outcome Prediction Using High-Order Brain Functional Connectivity Network

Xiuyi Jia[1,2], Han Zhang[2], Ehsan Adeli[2], and Dinggang Shen[2(✉)]

[1] School of Computer Science and Engineering, Nanjing University of Science and Technology, Nanjing, China
[2] Department of Radiology and BRIC, UNC at Chapel Hill, Chapel Hill, NC, USA
dgshen@med.unc.edu

Abstract. Based on the neuroimaging data from a large set of acquired brain injury patients, we investigate the feasibility of using machine learning for automatic prediction of individual consciousness level. Rather than using the traditional Pearson's correlation-based brain functional network, which measures only the simple temporal synchronization of the BOLD signals from each pair of brain regions, we construct a high-order brain functional network that is capable of characterizing topographical information-based high-level functional associations among brain regions. In such a high-order brain network, each node represents the community of a brain region, described by a set of this region's low-order functional associations with other brain regions, and each edge characterizes topographical similarity between a pair of such communities. Experimental results show that the high-order brain functional network enables a significant better classification for consciousness level and recovery outcome prediction.

1 Introduction

Studying the relationship between consciousness and brain activity has drawn a lot of attention in the recent years, especially using resting-state functional MRI (rs-fMRI) to investigate how brain functional network supports consciousness [1, 2]. The resting-state brain functional architecture can be characterized by different brain networks defined by correlated spontaneous brain activity between the regions of interest (ROIs). However, it is still unclear which key brain regions and their corresponding networks are essential to consciousness emergence and maintenance [3]. Perri *et al.* [4] reported that negative default mode network (DMN) connectivity seemed to be of metabolic neuronal origin, characterized by patients who have emerged from disorders of consciousness. Qin *et al.* [5] investigated three different functional networks to distinguish between conscious and unconscious states, and found that the salience network connectivity correlated with consciousness, while the DMN connectivity can be used to predict the recovery of consciousness. Wu *et al.* [3] summarized that the functional connectivity strength mainly in the DMN was disrupted with varying degrees of consciousness loss, and hence this disruption could be a potential biomarker for consciousness level prediction.

© Springer International Publishing AG 2017
G. Wu et al. (Eds.): CNI 2017, LNCS 10511, pp. 17–24, 2017.
DOI: 10.1007/978-3-319-67159-8_3

In these previous works, brain networks were usually constructed first based on the simple Pearson's correlation (PC), and then a particular group-level statistical analysis, such as one-way ANalysis Of VAriance (ANOVA), were applied to investigate if there exists any significant group differences in the population-averaged brain networks between different consciousness-level groups. Note that the PC-based network construction only captures the pairwise relationships through simple correlation operations. It is incapable of capturing any higher-order, complex relations between the brain regions, thus causing difficulty for the subsequent statistical analyses to exploit the consciousness level. Moreover, the hypothesis-driven analysis, such as in [5], limits our understanding of the biological substrate of consciousness with respect to the whole-brain complex network due to simply including a few predefined brain regions while ignoring other brain regions' contribution. To address these limitations, we investigate the relevant machine learning methods for automatic prediction of individual consciousness level according to the whole-brain complex networks.

For the construction of complex brain networks, some previous research have utilized certain prior knowledge and network information for building the respective models. Typical models include sparse representation (SR) [6], joint low-rank and sparse (SLR) method [7], and weighted sparse group representation method [8]. However, again in all these models, the networks are constructed by considering only pairwise interactions between ROIs. The higher order relations between the ROIs (i.e., nodes in the brain network) were overlooked in most of the previous works. To extract the underlying complex relationships from the network, in this paper, we propose a simple but effective high-order brain functional connectivity network (BFCN) construction method. In particular, the high-order BFCN is constructed based on the conventional low-order BFCN. Each node in the high-order BFCN represents the community of each ROI described by a set of low-order network values, and the edge between each pair of the nodes represents the correlation between the two communities. This high-order BFCN can model complex interactions and relationships among brain regions at a higher level, without introducing any extra parameters.

We use our proposed high-order BFCN for prediction of individual consciousness level. Experimental results on using rs-fMRI data for acquired brain injury (ABI) classification show that the high-order network enables a successful classification between the consciousness preserved and unresponsive patients. We also apply our high-order BFCN to predict whether the unresponsive patients would regain consciousness, from which we obtain a promising accuracy of 87.18%.

2   Materials

Our dataset comprises 53 patients with ABI but with the fully preserved consciousness state, and 39 ABI patients with unresponsive wakefulness state (including 26 in vegetative state and 13 in coma). These different groups of patients are categorized as follows [9]. (1) The *preserved* consciousness patients were able to communicate and had experienced brain injury. (2) The *vegetative state* patients were characterized by no evidence of awareness of self or environment and also an inability to interact with others; no evidence of sustained, reproducible, purposeful, or voluntary behavioral responses to visual, auditory, tactile, or noxious stimuli; no evidence of language comprehension or expression; intermittent wakefulness manifested by the presence of sleep-wake cycles; sufficiently preserved hypothalamic and brainstem autonomic functions to permit survival with medical and nursing care; bowel and bladder incontinence; and variably preserved cranial nerve reflexes and spinal reflexes. (3) The *coma* patients were characterized by no arousal/eye-opening, no behavioral signs of awareness, impaired spontaneous breathing, impaired brainstem reflexes, and no vocalizations of more than 1 h. Both *vegetative state* and *coma* patients are categorized as "unresponsive" subjects, while all other patients belong to another group of "consciousness preserved" subjects. The rs-fMRI data of these patients were collected from 2010 to 2014 via a Siemens 3.0 T scanner with the following parameters: TR = 2 s, slice number = 33, slice thickness = 4 mm, matrix size = 64 × 64. The data was preprocessed by using SPM8 (http://www.fil.ion.ucl.ac.uk/spm/) similar to [3]. It is important to note that the T1-weighted images of these subjects were also acquired and used to guide the registration using group-wise registration algorithm in SPM8 (DARTEL) for avoiding registration error due to lesions. The subjects with excessive head motion or large lesions that induced severe brain distortions were excluded during data screening.

The consciousness levels of the patients were assessed using the Glasgow Coma Scale (GCS) [10] and the Coma Recovery Scale-Revised (CRS-R) [11] on the day of the scanning. The recovery outcome was assessed using the Glasgow Outcome Scale (GOS) [12] at 3 months after scanning. The GOS provides a measurement of outcome, ranging from 1 to 5. The GOS score of less than 3 was defined as nonawakened, and the GOS score of larger or equal 3 as awakened [3]. In our 39 subjects (26 in vegetative state and 13 in coma), 17 of them regained consciousness after 3 months while the remained 22 of them were still nonawakened. We will learn a model with our high-order networks to predict both consciousness level (53 consciousness preserved *vs.* 39 unresponsive) and recovery outcome (17 awakened *vs.* 22 nonawakened).

3   High-Order BFCN Construction

Low-Order BFCN: In this subsection, we will introduce the basics of low-order BFCN construction method for brain disorder diagnosis, and then extend the definitions to capture high-order network characteristics in the next subsection.

Assume each brain is parcellated into N ROIs. Here, each ROI has a mean time series $x_i \in R^K$, $i = 1, 2, \ldots, N$, where K is the number of time points. x_i can be represented as $x_i = [x_{1i}; x_{2i}; \ldots; x_{Ki}]$. Thus, each subject is represented by a matrix, $X = [x_1, x_2, \ldots, x_N] \in R^{K \times N}$. The BFCN construction is simply defined as finding a connectivity matrix $W \in R^{N \times N}$, which can be formulated as a matrix-regularized network learning method [7]:

$$\min_W f(X, W) + \lambda R(W) \tag{1}$$

where $f(X, W)$ is a data-fitting term, and $R(W)$ is a matrix-regularized term. Using different $f(X, W)$ or $R(W)$, we can obtain different BFCN construction methods. For instance, in the Pearson's correlation (PC) coefficient-based BFCN, the connectivity matrix is calculated by [13]:

$$\min_W \sum_{i,j=1}^{N} \left\| x_i - W_{ij} x_j \right\|^2, \tag{2}$$

Sparse representation (SR) is another popular method to construct BFCN:

$$\min_W \left(\sum_{i=1}^{N} \left\| x_i - \sum_{j \neq i} W_{ij} x_j \right\|^2 + \lambda \sum_{i=1}^{N} \sum_{j \neq i} \left| W_{ij} \right| \right), \tag{3}$$

where the regularization term enforces sparsity in the network, since it is known that BFCN is intrinsically sparse [14]. By importing the modularity prior as the matrix-regularized term, Qiao $et\ al.$ [7] also proposed joint sparsity and low rank (SLR) regularizations in BFCN construction, by using both L_1-norm and trace norm of W:

$$\min_W \left(\sum_{i=1}^{N} \left\| x_i - \sum_{j \neq i} W_{ij} x_j \right\|^2 + \lambda_1 \| W \|_1 + \lambda_2 \| W \|_* \right). \tag{4}$$

Note that if we set $\lambda_1 = 0$, we would have only a low-rank regularization. As can be seen, all these BFCN construction methods use pairwise relationships between ROIs, and hence they are low-order network construction techniques.

High-Order BFCN: We propose a high-order BFCN construction method, which can implicitly capture high-order relationships among ROIs, rather than just the pairwise relations. Specifically, we propose to capture a second-level relationship built on the previous lower-order BFCN. As a result, we can additionally capture inter-regional resemblances in the BFCN. In order to achieve this goal, we can first use any method for constructing the low-order BFCNs as introduced in the previous subsection. In this low-order network, for each node (i.e., brain ROI), we have a vector (i.e., rows in the low-order network matrix) measuring the relations between this node and all other nodes. Let's call this vector a node's *community*. Then, based on this low-order BFCN, a second layer of correlations can be computed between any pairs of brain ROIs. Figure 1 illustrates the computation procedure to build a high-order network, given a low-order BFCN.

Specifically, assume $W_j = \left[W_{1j}, \ldots, W_{mj}, \ldots, W_{Nj} \right]$ represent the community of the node j (corresponding to the ROI x_j) described by a set of $W_{mj}, \forall m \in \{1..N\}$. Here, W_{mj} represents the interaction relationship of the node m and the node j. Thus, we can calculate the Pearson's correlation coefficients between the node j's community and any arbitrary node q's community as follows:

Low-Order Network   High-Order Network

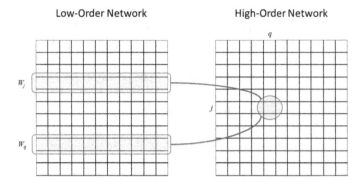

Fig. 1. Construction of high-order BFCN based on low-order BFCN. Each element in the high-order BFCN is calculated based on a pair of ROI communities from the low-order BFCN.

$$H_{jq} = \frac{\left(W_j - \overline{W_j}\right)^T \left(W_q - \overline{W_q}\right)}{\sqrt{\left(W_j - \overline{W_j}\right)^T \left(W_j - \overline{W_j}\right)} \sqrt{\left(W_q - \overline{W_q}\right)^T \left(W_q - \overline{W_q}\right)}}. \tag{5}$$

This way, H_{jq} would be a correlation between the communities of the two nodes j and q. Hence, it describes a more complex relationship between ROI x_j and ROI x_q at a higher level. With the assumption that W_j has been centralized by $W_j - \overline{W_j}$ and further normalized by $\sqrt{\left(W_j - \overline{W_j}\right)^T (W_j - \overline{W_j})}$, the PC coefficient can be simply represented as $H_{jq} = W_j^T W_q$. In the high-order network, the new edge between nodes j and q would have the weight of H_{jq}. Dropping the indices and writing in a matrix form, we would have $H = W^T W$ to represent the high-order BFCN. Under such settings, it is easy to construct high-order networks from the corresponding various low-order networks, such as $H_{(PC)} = W_{(PC)}^T W_{(PC)}$ and $H_{(SLR)} = W_{(SLR)}^T W_{(SLR)}$, with $W_{(PC)}$ and $W_{(SLR)}$ as the low-order BFCNs estimated based on *Pearson's correlation* and *sparse and low-rank regularization*, respectively.

It is worth pointing out that some machine learning methods also tried to use the linear transformation $W^T W$ to select features [15, 16]. The main difference is that these machine learning methods aim at solving the over-determined problem (with more subjects than features in matrix) or under-determined problem (with more features than subjects in W) by using the linear transformation. In our work, we do not have these problems as our network W is a $N \times N$ matrix, and we want to use the high-order network $W^T W$ to extract more complex correlation on community level.

4   Experiments

Network Construction and Experimental Setting: In our experiments, for each subject, 200 ROIs are defined based on Craddock's 200 atlas, and the mean rs-fMRI signals are extracted from each ROI to construct the BFCN. We construct two types of low-order BFCNs as described in Sect. 2, including PC and SLR. Based on these two low-order BFCNs, we can construct two respective high-order BFCNs, namely H_PC and H_SLR. For the regularization tuning parameters (i.e., λ_1 and λ_2 in Eq. 4) involved in the SLR low-order BFCN model, we use the same setting as in [7], and search their optimal values in the set $\{2^{-5}, 2^{-4}, \ldots, 2^0, \ldots, 2^4, 2^5\}$. Note that there are no parameters to tune for the construction of high-order BFCN.

As we have 200 ROIs as nodes in a network, and since the connectivity matrix is symmetric, we will vectorize the lower-triangle of the matrix and use it as the feature vector for each network. As a result, we will have $\dfrac{200 \times (200 - 1)}{2} = 19900$ edges to describe each connectivity network. For each of the BFCNs, we use these edge strength values in the networks as features. We then select the most informative features among all these features, and then learn a classifier model. For feature selection, we use a simple information theoretic feature selection techniques, which evaluates the information gain for every single feature, and selects the features with the most information gain. Specifically, we measure the information gain ratio with respect to the class label for each feature, similar to [17]. Then, we choose all the features with information gain ratio values larger than 0.01 in our experiments. After the selected features are identified, we employ a polynomial kernel SVM with $c = 1$ as the classifier.

Classification Results on Consciousness Level Prediction: In the following, we report four evaluation measures: accuracy, sensitivity, specificity, and F-Score for both low-order and high-order networks using the above feature selection and classification methods. The reported values are the mean of 10 different runs of 10-fold cross validation, and hence introduce reliable results with no over-fitting effects to the particular population of the data. For selection of tuning parameters in SLR, we further conduct an inner leave-one-out cross validation on the training set to obtain their best parameter values.

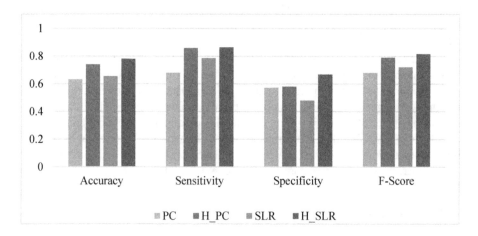

Fig. 2. Comparison of classification results between low-order BFCNs (including PC and SLR) and high-order BFCNs (including H_PC and H_SLR).

The results are shown in Fig. 2. From these results, we can conclude that the high-order BCNF obtains a better performance in all experiments. Furthermore, we can see that H_SLR obtains the best classification results, with 78.04% accuracy, 86.39% sensitivity, 66.82% specificity, and 81.72% F-Score.

Results on Recovery Outcome Prediction: From the previous subsection, we can see that the H_SLR can generate the best classification accuracy results. Therefore, we use H_SLR to predict the recovery outcome of unresponsive patients. In order to be able to compare with previous methods [3, 5] on the same application, we implement two different cross-validation settings: (1) leave-one-out cross validation (LOOCV), and (2) 5 runs of leave-two-out cross validation (LTOCV). Note that, for the 2nd case, we average results from 5 different runs and reported the average.

As shown in Table 1, our proposed method can obtain the best accuracy of 87.18%, compared to other methods, under LOOCV.

Table 1. Comparison of accuracy on different methods.

Method	LOOCV (%)	LTOCV (%)
H_SLR +SVM	87.18	81.54
PC + SVM [3]	81.25	75.61
PC + ANOVA [5]	74.00*	N/A

Note that the comparisons are conducted under different cross-validation mechanism. N/A means the result is not available from the corresponding reference. * This accuracy was obtained based on a classification model trained using all subjects (i.e., not via stringent machine learning).

5 Conclusion

In this paper, we proposed a simple *but* effective high-order brain functional connectivity network construction method for predicting both consciousness level and recovery outcome in acquired brain injury. Our proposed high-order network treats the community of each ROI as its features and the correlation between any pair of communities as the edge between the two ROIs. Compared to the low-order network, the high-order network can extract more information at the high level. The experiments on both consciousness level prediction and recovery outcome prediction in ABI show that our proposed high-order network can obtain a better classification performance. In future work, other relationships between communities will also be investigated to build the high-level network.

Acknowledgements. This work is partially supported by National Natural Science Foundation of China (Grant Nos. 61403200), Natural Science Foundation of Jiangsu Province (Grant No. BK20140800), and NIH grants (EB006733, EB008374, EB009634, MH107815, AG041721, and AG042599).

References

1. Sharp, D.J., Scott, G., Leech, R.: Network dysfunction after traumatic brain injury. Nat. Rev. Neurol. **10**, 156–166 (2014)
2. Shulman, R.G., Hyder, F., Rothman, D.L.: Baseline brain energy supports the state of consciousness. Proc. Natl. Acad. Sci. U.S.A. **106**, 11096–11101 (2009)
3. Wu, X., Zou, Q., Hu, J., et al.: Intrinsic functional connectivity patterns predict consciousness level and recovery outcome in acquired brain injury. J. Neurosci. **35**(37), 12932–12946 (2015)
4. Perri, C.D., Bahri, M.A., Amico, E.: Neural correlates of consciousness in patients who have emerged from a minimally conscious state: a cross-sectional multimodal imaging study. Lancet Neurol. **15**, 830–842 (2016)
5. Qin, P., Wu, X., Huang, Z., et al.: How are different neural networks related to consciousness? Ann. Neurol. **78**, 594–605 (2015)
6. Huang, S., Li, J., Sun, L., et al.: Learning brain connectivity of Alzheimer's disease from neuroimaging data. In: Bengio, Y., et al. (eds.): NIPS 2009, pp. 808–816 (2009)
7. Qiao, L., Zhang, H., Kim, M., et al.: Estimating functional brain networks by incorporating a modularity prior. Neuroimage **141**, 399–407 (2016)
8. Yu, R., Zhang, H., An, L., Chen, X., Wei, Z., Shen, D.: Correlation-weighted sparse group representation for brain network construction in mci classification. In: Ourselin, S., Joskowicz, L., Sabuncu, Mert R., Unal, G., Wells, W. (eds.) MICCAI 2016. LNCS, vol. 9900, pp. 37–45. Springer, Cham (2016). doi:10.1007/978-3-319-46720-7_5
9. Schnakers, C.: Clinical assessment of patients with disorders of consciousness. Arch. Ital. Biol. **150**, 36–43 (2012)
10. Teasdale, G., Jennett, B.: Assessment of coma and impaired consciousness. Pract. Scale Lancet **2**, 81–84 (1974)
11. Giacino, J.T., Kalmar, K., Whyte, J.: The JFK coma recovery scale-revised: measurement characteristics and diagnostic utility. Arch. Phys. Med. Rehabil. **85**, 2020–2029 (2004)
12. Jennett, B., Bond, M.: Assessment of outcome after severe brain damage. Lancet **1**, 480–484 (1975)
13. Lee, H., Lee, D.S., Kang, H., et al.: Sparse brain network recovery under compressed sensing. IEEE Trans. Med. Imaging **30**, 1154–1165 (2011)
14. Sporns, O.: Networks of the Brain. MIT Press, Cambridge (2011)
15. Lanckriet, G.R.G., Cristianini, N., Bartlett, P., et al.: Learning the kernel matrix with semidefinite programming. J. Mach. Learn. Res. **5**, 27–72 (2004)
16. Munsell, B.C., Wee, C.Y., Keller, S.S., et al.: Evaluation of machine learning algorithms for treatment outcome prediction in patients with epilepsy based on structural connectome data. Neuroimage **118**, 219–230 (2015)
17. Karegowda1, A.G., Manjunath, A.S., Jayaram, M.A.: Comparative study of attribute selection using gain ratio and correlation based feature selection. Int. J. Inf. Technol. Knowl. **22**(2), 271–277 (2010)

Discriminative Log-Euclidean Kernels for Learning on Brain Networks

Jonathan Young[1(✉)], Du Lei[2], and Andrea Mechelli[2]

[1] Centre for Neuroimaging Sciences, Institute of Psychiatry,
Psychology and Neuroscience, King's College London, London, UK
jonathan.young@kcl.ac.uk
[2] Department of Psychosis Studies, Institute of Psychiatry,
Psychology and Neuroscience, King's College London, London, UK

Abstract. The increasing availability of functional Magnetic Resonance Imaging (fMRI) has led to a number of studies of brain networks with the aim of developing computer aided diagnosis of disease. Typically these are based on a statistical or machine learning method operating on connectivity networks, or features derived from them. This work presents a novel kernel method allowing classification tasks on connectivity networks represented as symmetric positive definite (SPD) matrices. It defines a kernel based on geodesic distances measured on the Riemannian manifold of SPD matrices, and automatically adjusts the eigenvalues of the matrices to improve accuracy. This is coupled with a Gaussian Process (GP) classifier, and used to discriminate healthy controls from Schizophrenia patients. The new kernel offers superior classification accuracy to previous kernels, and the adjusted eigenvalues allow discovery of clinically meaningful differences in connectivity between patients and controls.

1 Introduction

Brain networks based on resting state functional magnetic resonance imaging (rs-fMRI) have become increasingly prominent in the study and, potentially, diagnosis of conditions affecting neural functioning. The brain is represented as a graph whose nodes correspond to anatomical regions defined by an atlas, while edges are weighted with the estimated functional connectivity between pairs of regions. The edge weights can be calculated as the simple correlation between the fMRI signal timecourses of a pair of regions. However sparse inverse covariance matrix estimation, which estimates partial correlations to remove the effect of indirect connections, is more sensitive [14]. The resulting connectivity graphs can be conveniently represented as symmetric positive definite (SPD) adjacency matrices.

Comparing brain networks between subjects then becomes a problem of comparing SPD matrices. These do not lie in a Euclidean space, but rather form a Riemannian manifold. A number of recent publications have used operations on this manifold, either to manipulate the matrices to improve discrimination

© Springer International Publishing AG 2017
G. Wu et al. (Eds.): CNI 2017, LNCS 10511, pp. 25–34, 2017.
DOI: 10.1007/978-3-319-67159-8_4

between clinical groups [8] or by defining a kernel function on the manifold that can be used directly in classification with a kernel method such as a support vector machine (SVM) [3,4]. Kernel functions on the manifold are based on the Gaussian radial basis function (RBF) [5]. In the Euclidean space $\mathbb{R}^n$, this is given by $k(\mathbf{x}, \mathbf{y}) = \exp(-\gamma \|\mathbf{x} - \mathbf{y}\|^2)$ for a pair of n dimensional vectors. On a manifold, the Euclidean distance is replaced by a distance induced by the Riemannian metric. A number of metrics have been proposed for the manifold of SPD matrices, $\mathbf{Sym}^+$. However many of these suffer from various drawbacks for use as kernel functions in machine learning [5]. The widely used affine invariant metric [10] induces a true geodesic distance, but is computationally expensive and can result in a non positive definite Gaussian kernel for some values of γ. The Stein divergence [15] is fast, but does not induce a geodesic distance and gives a positive definite RBF kernel only for a non-continuous range of values of γ, which poses difficulties for parameter tuning. The log-Euclidean metric [1], on the other hand, suffers from neither difficulty.

Although using kernels based on distances under metrics such as the log-Euclidean is an elegant and powerful approach, there are two deficiencies where improvements can potentially be made. Firstly, the estimated eigenvalues may be biased or inaccurate if the sample size is small. Secondly, the eigenvalues may not be optimal for the desired purpose of discriminating between two clinical groups. A method to tackle these issues is proposed in [17]. The authors use a kernel based on the Stein divergence due to its computational efficiency, and adjust the eigenvalues of SPD matrices by learning a set of coefficients for their original eigenvalues. This is done by performing an optimisation of functions such as kernel alignment. The adjusted eigenvalues provide improvement in accuracy in a variety of tasks including brain network classification.

Our work presented here instead makes use of the log-Euclidean kernel, selected for the reasons described above. The eigenvalues of the original SPD matrices are multiplied by a set of coefficients learned from the training data, but rather than optimising kernel alignment we maximise the log likelihood of the data under a Gaussian process framework. This has a number of advantages. Chiefly, the eigenvalue coefficients and the length scale parameter γ can be optimised jointly rather than separately any without any need for cross-validation and the Gaussian process framework provides a great deal of flexibility in terms of what types of learning problems can be addressed. Although here we restrict ourselves to binary classification, this can be extended to multiclass classification, regression, and time to event analysis via different likelihood functions. We show that this provides improved accuracy in classifying Schizophrenia patients and controls in a publically available dataset, and that the adjusted eigenvalues yield matrices that show significant differences in connectivity between the two clinical groups.

2   Methods and Materials

2.1   Subjects

The methods are applied to at set of 100 subjects drawn from the publically available COBRE dataset, described in [2]. This consists of 53 patients and 47 healthy controls. Schizophrenia was diagnosed using the Structured Clinical Interview for DSM Disorders. Subjects with schizophrenia were excluded if they had a DSM-IV Axis I diagnosis of other neurological disorders, mental retardation, if they had suffered severe head trauma with a loss of consciousness greater than 5 min loss of consciousness, or if they had history of substance abuse or dependence during the last 12 months. Healthy control subjects were excluded if they had any DSM-IV Axis I mental disorders, other neurological conditions, or a history of substance abuse or head trauma.

2.2   Image Data

All subjects were scanned on a 3-Tesla Siemens Trio scanner with a 12-channel head coil. The T1-weighted structural images used to define anatomical regions in each subject were acquired with a five echo MEMPR sequence with TE = 1.64, 3.5, 5.36, 7.22, and 9.08 ms, TR = 2.53, and voxel size = 1 mm isotropic. The resting-state fMRI images were acquired with a gradient-echo EPI sequence with TE = 29 ms, TR = 2 s, and a voxel size $3 \times 3 \times 4$ mm. For more detailed information on this dataset, see[1].

2.3   Image Processing

The preprocessing of rs-fMRI data was performed as follows. SPM8[2] was used to perform image preprocessing. The first 10 time points were discarded to avoid instability of the initial MRI signal. All the time points after 150 were also discarded. Then, the images were corrected for intra-volume acquisition time delay and inter-volume geometric displacement of head motion. After these corrections, the images were spatially normalized to a $3 \times 3 \times 3$ mm^3 Montreal Neurological Institute (MNI) 152 template and then linearly detrended and temporally band-pass filtered (0.01–0.08 Hz) to remove low-frequency drift and high-frequency physiological noise. Finally the global signal, the white matter signal, the cerebrospinal fluid (CSF) signal and the motion parameters (1.5 mm translational and 1.5° rotational parameters) were regressed out.

2.4   Brain Network Construction

Networks were constructed using GRETNA software[3]. First, the whole brain was divided into 90 cortical and subcortical regions of interest using the automated

---

[1] tinyurl.com/fcon1000-cobre.

[2] http://www.fil.ion.ucl.ac.uk/spm.

[3] http://www.nitrc.org/projects/gretna/.

anatomical labelling (AAL) atlas [16] with each region representing a network node. Then, the mean time series of each region were generated. As a final step, sparse inverse covariance matrix estimation was applied to the set of time series for each subject. The regularisation parameter was set by cross-validation. The resulting precision matrices were inverted to provide an SPD covariance matrix representing the brain network for each subject.

2.5   Gaussian Process Classification

Before introducing the discriminative log-Euclidean kernel, it is necessary to briefly discuss the Gaussian process (GP) classification we use it with in our experiments. For a more detailed treatment see [12]. Formally speaking, a GP is a generalisation of ordinary multivariate Gaussian distributions to the case of an infinite number of variables, which means it can be seen as a distribution over functions. In practice, any finite subset of the variables forms a multivariate Gaussian. Hence a Gaussian process, like a multivariate Gaussian, can be fully parameterised by a mean vector $\mathbf{m}$ and a covariance matrix $\mathbf{K}$. In a GP, the elements of these are determined a mean function $m(\mathbf{x}, \boldsymbol{\theta})$ and a covariance function $k(\mathbf{x}, \mathbf{x}', \boldsymbol{\theta})$, where $\mathbf{x}$ and $\mathbf{x}'$ are data vectors and $\boldsymbol{\theta}$ is a vector collecting hyperparameters which control the functional form of the mean and covariance functions. So we can write

$$GP \sim \mathcal{N}(\mathbf{m}, \mathbf{K}) \tag{1}$$

where

$$\mathbf{m} = \begin{bmatrix} m(\mathbf{x}) \\ m(\mathbf{x}') \end{bmatrix}, \ \mathbf{K} = \begin{bmatrix} k(\mathbf{x}, \mathbf{x}) \ k(\mathbf{x}, \mathbf{x}') \\ k(\mathbf{x}', \mathbf{x}) \ k(\mathbf{x}', \mathbf{x}') \end{bmatrix} \tag{2}$$

In Gaussian processes for regression, the GP forms a prior over the space of regression functions $f(\mathbf{x})$ which map a data vector $\mathbf{x}$ onto a regression target value y. Assuming a (univariate) Gaussian likelihood function with variance σ^2, and for simplicity a zero mean function, we can then model the regression as

$$\begin{aligned} y_i &= f(\mathbf{x}_i) + \epsilon \\ f &\sim \mathcal{N}(\boldsymbol{\mu} = 0, \mathbf{K}) \\ \epsilon &\sim \mathcal{N}(0, \sigma^2) \end{aligned} \tag{3}$$

By then applying Bayes' rule, we can calculate the resulting posterior, which gives a predictive distribution for an unseen test vector $\mathbf{x}^*$. For a set of training vectors $\mathbf{X}$ and corresponding training targets $\mathbf{y}$, the predictive distribution for the value of $f(\mathbf{x}^*)$ is given by

$$\begin{aligned} p(f(\mathbf{x}^*)|\mathbf{X}, \mathbf{y}, \mathbf{x}^*, \boldsymbol{\theta}) &\sim \mathcal{N}(\mu^*, \sigma^{2*}) \\ \mu^* &= \mathbf{k}^{*\top} \mathbf{C}^{-1} \mathbf{y} \\ \sigma^{2*} &= k(\mathbf{x}^*, \mathbf{x}^*) = \mathbf{k}^{*\top} \mathbf{C}^{-1} \mathbf{k}^* \end{aligned} \tag{4}$$

where $\mathbf{C} = \mathbf{K} + \sigma^2\mathbf{I}$. $\mathbf{K}$ is the covariance matrix derived from the covariance function k, training data $\mathbf{X}$ and covariance hyperparameters $\boldsymbol{\theta}$, so $\mathbf{K}_{i,j} = k(\mathbf{x}_i, \mathbf{x}_j, \boldsymbol{\theta})$, and $\mathbf{k}^*$ is a vector of covariances between the test data point $\mathbf{x}^*$ and all the training data points.

Binary classification is more complex, as the Gaussian likelihood function is replaced with a sigmoid, mapping the latent function $f(\mathbf{x})$ to interval $[0, 1]$, representing the probability a data vector having a particular class. This results in the posterior being non-Gaussian, and hence they must be approximated. Here we make use of expectation propagation (EP) [7], as this has been shown to provide results as accurate as the gold standard Monte-Carlo Markov Chain (MCMC) methods for this task, while being far faster to compute [9].

The GP covariance function hyperparameters $\boldsymbol{\theta}$ control its behaviour. For the discriminative log-Euclidean kernel these are comprised of the vector of eigenvalue coefficients $\mathbf{a}$ and characteristic length scale γ for our kernel function. These are tuned by type-II maximum likelihood with gradient decent, making use of the derivatives from Eq. 10.

2.6 The Discriminative Log-Euclidean Kernel

The conventional log-Euclidean kernel is based on the log-Euclidean metric introduced in [1]. The log-Euclidean geodesic distance between SPD matrices $\mathbf{S}_1$ and $\mathbf{S}_2$ is given by

$$d_{LE}(\mathbf{S}_1, \mathbf{S}_2) = \|\mathrm{logm}(\mathbf{S}_1) - \mathrm{logm}(\mathbf{S}_2)\|_F \qquad (5)$$

where logm is the matrix logarithm and $\|\cdot\|_F$ the Frobenius norm. Hence, the RBF log-Euclidean kernel is

$$k_{LE}(\mathbf{S}_1, \mathbf{S}_2) = \exp\left(\frac{-d_{LE}(\mathbf{S}_1, \mathbf{S}_2)^2}{\gamma^2}\right) \qquad (6)$$

The functional form of the discriminative RBF log-Euclidean kernel is identical, but the original matrices $\mathbf{S}$ are replaced by adjusted matrices $\mathbf{S}^*$. The adjustment is done by rescaling the eigenvalues of $\mathbf{S}$, by multiplying them by a vector of d coefficients $\mathbf{a}$, where d is the dimensionality of the matrices. Using the eigen decomposition of SPD matrices, if $\mathbf{S} = \mathbf{U}\boldsymbol{\Lambda}\mathbf{U}^\top$, where $\mathbf{U}$ is a matrix whose columns are the eigenvectors and $\boldsymbol{\Lambda}$ is a diagonal matrix of the eigenvalues $(\lambda_1, ..., \lambda_d)$, then

$$\mathbf{S}^* = \mathbf{U}\begin{pmatrix} \lambda_1\mathbf{a}_1 & & & \\ & \lambda_2\mathbf{a}_2 & & \\ & & \ddots & \\ & & & \lambda_d\mathbf{a}_d \end{pmatrix}\mathbf{U}^\top \qquad (7)$$

and

$$\text{logm}(\mathbf{S}^*) = \mathbf{U} \begin{pmatrix} \log(\lambda_1 \mathbf{a}_1) & & & \\ & \log(\lambda_2 \mathbf{a}_2) & & \\ & & \ddots & \\ & & & \log(\lambda_d \mathbf{a}_d) \end{pmatrix} \mathbf{U}^\top \tag{8}$$

We want to set the values of $\mathbf{a}$ with gradient based optimisation. From Eq. 8 we can see that $\text{logm}(\mathbf{S}^*)_{i,j} = \sum_{k=1}^d \mathbf{U}_{i,k}\mathbf{U}_{j,k}\log(\lambda_k \mathbf{a}_k)$. Hence the derivatives of $\text{logm}(\mathbf{S}^*)$ with respect to the kth value of $\mathbf{a}$ are given by

$$\frac{\partial}{\partial \mathbf{a}_k}\text{logm}(\mathbf{S}^*)_{i,j} = \frac{\mathbf{U}_{i,k}\mathbf{U}_{j,k}}{\mathbf{a}_k}. \tag{9}$$

Now differentiating Eq. 6 with respect to $\mathbf{a}_k$, we obtain

$$\begin{aligned} \frac{\partial}{\partial \mathbf{a}_k} k_{LE}(\mathbf{S}_1^*, \mathbf{S}_2^*) &= k_{LE}(\mathbf{S}_1^*, \mathbf{S}_2^*)\frac{\partial}{\partial \mathbf{a}_k}\frac{-d_{LE}(\mathbf{S}_1^*, \mathbf{S}_2^*)^2}{\gamma^2} \\ &= k_{LE}(\mathbf{S}_1^*, \mathbf{S}_2^*)\frac{-2}{\gamma^2}\mathbf{1}^\top s_{LE}(\mathbf{S}_1^*, \mathbf{S}_2^*) \odot \frac{\partial}{\partial \mathbf{a}_k}s_{LE}(\mathbf{S}_1^*, \mathbf{S}_2^*)\mathbf{1} \end{aligned} \tag{10}$$

where $\mathbf{1}$ is a length d vector of ones, $\odot$ is the Hadamard or elementwise product, and $s_{LE}(\mathbf{S}_1^*, \mathbf{S}_2^*)$ is $\text{logm}(\mathbf{S}_1^*) - \text{logm}(\mathbf{S}_2^*)$. By plugging Eq. 9 into Eq. 10 we obtain a formula for the derivatives of the kernel functions with respect to the elements of the vector of eigenvalue coefficients $\mathbf{a}$.

2.7   Impementation Details

The discriminative log-Euclidean kernel is implemented as a new covariance function for the GPML toolbox[4]. The values of $\mathbf{a}$ are optimised in the log domain to keep $\mathbf{a}$ positive. To avoid repeating the computationally costly eigendecomposition, these are precomputed for all SPD connectivity matrices. The covariance function then operates on vectors consisting of the concatenated eigenvalues and eigenvectors of a matrix. The values of $\mathbf{a}$ are initialised to 1 and the initial value of γ is the median d_{LE} between all pairs of matrices in the training data.

2.8   Classification Experiments

For the classification experiments, a Monte-Carlo cross validation (MCCV) procedure was used, where the data and labels were randomly shuffled 200 times. After each shuffle, the first 90% of matrices and labels were used for training and the remaining 10% for testing. The results of all 200 test sets were then averaged. For comparison, the same procedure is used with the conventional log-Euclidean kernel (i.e., unadjusted eigenvalues), and with a linear kernel and Gaussian kernel on vectors of the elements of each connectivity matrix (i.e., assuming the connectivities lie in a Euclidean space).

---

[4] http://www.gaussianprocess.org/gpml/code/matlab/doc/.

2.9 Group Difference Experiments

To examine the possibility of finding group differences in connectivity with the adjusted eigenvalues, a tenfold cross-validation experiment was performed. For each fold, 90 subjects were used for training, yielding a set of values for the eigenvalue coefficients **a**. These coefficients were then used to adjust the eigenvalues in the connectivity matrices for the 10 subjects set aside for testing. The process was repeated so that eigenvalues of all 100 subjects were adjusted. To calculate the statistical significance of differences in connectivity between patients and controls, for each matrix element in the upper triangle of the connectivity matrix, the connectivity values for all subjects in the matrix element were extracted, and a two sided t-test was applied to the values. Finally, we corrected for multiple comparisons with false discovery rate correction.

3 Results and Discussion

Results of classification are shown in Table 1. The discriminative log-Euclidean kernel produces higher accuracy than the alternative kernels, with an improvement of nearly 4% over the basic linear kernel. Surprisingly, perhaps, the difference gained by operating on the manifold is small, although both manifold kernels outperform the Euclidean ones. Sensitivity and specificity are reasonably well balanced at 65.7% and 82.6% respectively.

The adjustments to the eigenvalues tend to be small, with values of **a** close to unity. The largest adjustments are mostly negative (shrinking the eigenvalue) and are associated with the largest eigenvalues. This supports the idea that the larger eigenvalues can be overestimated [17].

Table 1. Results of classification experiments

	Euclidean kernels		Manifold kernels	
	Linear	RBF	Log-Euclidean	Discriminative log-Euclidean
Accuracy (%)	71.35	72.30	72.85	**74.95**

The pattern of significant differences in connectivity between patients and controls in the adjusted matrices is shown in Fig. 1. p-values are inverted in the figure so brighter colours mean a higher level of significance.

Brain regions found to show the most significant differences in functional connectivity, using both the original and adjusted connectivity matrices, include the temporal gyri and striatum with stronger connectivity in patients than controls, and reduced connection strength in patients in both left and right fusiform gyri. Several previous studies have reported functional dysconnectivity in these regions in Schizophrenia patients relative to controls (see [11] for review). Furthermore, functional dysconnectivity in these regions has been reported to

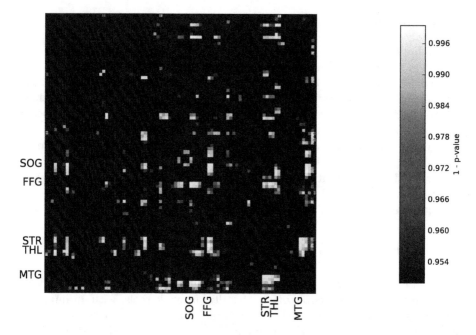

Fig. 1. Connections with significant difference ($p < 0.05$) between patients and controls using the eigenvalue-adjusted connectivity matrices, according a 2-sided test. Brain regions with clusters of significant connectivity are labelled: SOG = L and R superior occipital gyri, FFG = L and R fusiform gyri, STR = striatum, THL = thalamus, MTG = L and R mediotemporal gyri.

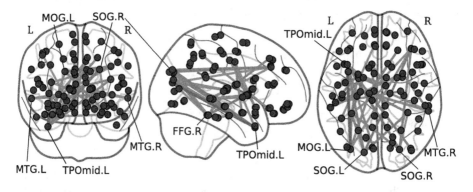

Fig. 2. Connections between atlas regions with most significant differences between patients and controls. Hubs of significant connectivity differences are labelled according to their AAL abbreviations. Visualisations from Nilearn, http://nilearn.github.io/.

depend on the symptom profile of patients [6,11,13]. Overall, the differences in connectivity identified with our technique are consistent with the existing literature. The anatomical patterns of significant connectivity are shown in Fig. 2. For clarity, only the most significantly different ($p < 0.005$) connections are shown.

The primary limitation of this technique is that it is restricted to operating on SPD matrices, which are not provided by all types of connectivity, especially structural connectivity obtained from tractography. However it should be possible to circumvent this by using graph Laplacians as in [4].

4   Conclusions

We have introduced the discriminative log-Euclidean kernel, and shown that by allowing the eigenvalues of matrices representing brain connectivity networks to be adjusted, it can provide superior classification ability to previous kernels, including one operating on the manifold of SPD matrices. Although the resulting changes to the eigenvalues alter the connectivity matrices, the ability to discover clinically important differences in connectivity between regions from the matrices is unaffected.

References

1. Arsigny, V., Fillard, P., Pennec, X., Ayache, N.: Fast and simple computations on tensors with log-euclidean metrics. Research report RR-5584, INRIA (2005)
2. Çetin, M.S., Christensen, F., Abbott, C.C., Stephen, J.M., Mayer, A.R., Caive, J.M., Bustillo, J.R., Pearlson, G.D., Calhoun, V.D.: Thalamus and posterior temporal lobe show greater inter-network connectivity at rest and across sensory paradigms in schizophrenia. Neuroimage **97**, 117–126 (2014)
3. Dodero, L., Minh, H.Q., San Biagio, M., Murino, V., Sona, D.: Kernel-based classification for brain connectivity graphs on the Riemannian manifold of positive definite matrices. In: 2015 IEEE 12th International Symposium on Biomedical Imaging (ISBI), pp. 42–45, April 2015
4. Dodero, L., Sambataro, F., Murino, V., Sona, D.: Kernel-based analysis of functional brain connectivity on grassmann manifold. In: Navab, N., Hornegger, J., Wells, W.M., Frangi, A.F. (eds.) MICCAI 2015. LNCS, vol. 9351, pp. 604–611. Springer, Cham (2015). doi:10.1007/978-3-319-24574-4_72
5. Jayasumana, S., Hartley, R., Salzmann, M., Li, H., Harandi, M.: Kernel methods on riemannian manifolds with gaussian RBF kernels. IEEE Trans. Pattern Anal. Mach. Intell. **37**(12), 2464–2477 (2015)
6. Mechelli, A., Allen, P., Amaro, E., Fu, C.H.Y., Williams, S.C.R., Brammer, M.J., Johns, L.C., McGuire, P.K.: Misattribution of speech and impaired connectivity in patients with auditory verbal hallucinations. Hum. Brain Mapp. **28**(11), 1213–1222 (2007)
7. Minka, T.P.: Expectation propagation for approximate Bayesian inference. In: Proceedings of the 17th Conference in Uncertainty in Artificial Intelligence (UAI 2001), pp. 362–369. Morgan Kaufmann Publishers Inc. (2001)
8. Ng, B., Dressler, M., Varoquaux, G., Poline, J.B., Greicius, M., Thirion, B.: Transport on riemannian manifold for functional connectivity-based classification. In: Golland, P., Hata, N., Barillot, C., Hornegger, J., Howe, R. (eds.) MICCAI 2014. LNCS, vol. 8674, pp. 405–412. Springer, Cham (2014). doi:10.1007/978-3-319-10470-6_51
9. Nickisch, H., Rasmussen, C.: Approximations for binary gaussian process classification. J. Mach. Learn. Res. **9**, 2035–2078 (2008)

10. Pennec, X., Fillard, P., Ayache, N.: A riemannian framework for tensor computing. Int. J. Comput. Vis. **66**(1), 41–66 (2006)
11. Pettersson-Yeo, W., Allen, P., Benetti, S., McGuire, P., Mechelli, A.: Dysconnectivity in schizophrenia: where are we now? Neurosci. Biobehav. Rev. **35**(5), 1110–1124 (2011)
12. Rasmussen, C.E., Williams, C.K.I.: Gaussian Processes for Machine Learning. MIT Press, Cambridge (2006)
13. Sarpal, D.K., Robinson, D.G., Lencz, T., Argyelan, M., Ikuta, T., Karlsgodt, K., Gallego, J.A., Kane, J.M., Szeszko, P.R., Malhotra, A.K.: Antipsychotic treatment and functional connectivity of the striatum in first-episode schizophrenia. JAMA Psych. **72**(1), 5–13 (2015)
14. Smith, S.M., Miller, K.L., Salimi-Khorshidi, G., Webster, M., Beckmann, C.F., Nichols, T.E., Ramsey, J.D., Woolrich, M.W.: Network modelling methods for FMRI. Neuroimage **54**(2), 875–891 (2011)
15. Sra, S.: Positive definite matrices and the S-divergence, October 2011. arXiv:1110.1773 [math, stat]
16. Tzourio-Mazoyer, N., Landeau, B., Papathanassiou, D., Crivello, F., Etard, O., Delcroix, N., Mazoyer, B., Joliot, M.: Automated anatomical labeling of activations in SPM using a macroscopic anatomical parcellation of the MNI MRI single-subject brain. Neuroimage **15**(1), 273–289 (2002)
17. Zhang, J., Wang, L., Zhou, L., Li, W.: Learning discriminative stein kernel for SPD matrices and its applications. IEEE Trans. Neural Netw. Learn. Syst. **27**(5), 1020–1033 (2016)

Interactive Computation and Visualization of Structural Connectomes in Real-Time

Maxime Chamberland[1(✉)], William Gray[1], Maxime Descoteaux[2], and Derek K. Jones[1]

[1] CUBRIC, School of Psychology, Cardiff University, Cardiff, UK
ChamberlandM@cardiff.ac.uk
[2] SCIL, University of Sherbrooke, Sherbrooke, Canada

Abstract. Structural networks contain high dimensional data that raise huge computational and visualization problems, especially when attempting to characterise them using graph theory. As a result, it can be non-intuitive to grasp the contribution of each edge within a graph, both at a local and global scale. Here, we introduce a new platform that enables tractography-based networks to be explored in a highly interactive real-time fashion. The framework allows one to interactively tune graph-related parameters on the fly, as opposed to conventional visualization softwares that rely on pre-computed connectivity matrices. From a neurosurgical perspective, the method also provides enhanced understanding regarding the potential removal of a specific node or transection of an edge from the network, allowing surgeons and clinicians to discern the value of each node.

1 Introduction

The human brain can be viewed as a network [1]. This highly specialized network can be conceptualized to as a set of gray matter (GM) regions that are linked together by white matter (WM) connections, represented by graph nodes and edges respectively. Brain networks derived from graph theory analyses are often dense and complex, and thus perceptually challenging to visualize [11]. While thresholding edges can help reduce the complexity of a network, it often leads to high variance in graph metrics [6,8,13]. Moreover, false positive in tractography [4,10] pollute connectivity matrices and adversely impact on chosen graph metric.

To better understand the role of these confounding factors on network topology, we develop a new visualization framework for exploring structural networks in a highly-interactive fashion. More specifically, the proposed visualization framework: (1) provides real-time insight of various thresholds on graph metrics; and (2) enables a seamless transition between an graph abstract (nodes and edges) and an anatomical (streamlines) representation, allowing one to inspect the underlying architecture of a specific edge.

© Springer International Publishing AG 2017
G. Wu et al. (Eds.): CNI 2017, LNCS 10511, pp. 35–41, 2017.
DOI: 10.1007/978-3-319-67159-8_5

2   Methods

2.1   Structural Connectivity

Diffusion-weighted images of a single-subject were acquired along 64 uniformly-distributed directions at b = 1000 s/mm², using single-shot EPI on a 1.5 Tesla SIEMENS Magnetom (128 × 128 matrix, 2 mm isotropic resolution, TR/TE 11000/98 ms) and a GRAPPA factor of 2. An anatomical T1-weighted 1 mm isotropic MPRAGE (TR/TE 6.57/2.52 ms) image was also acquired for the estimation of partial volume maps (PVE). The diffusion-weighted images were upsampled to the anatomical resolution (1 mm isotropic). Fiber Orientation Distribution Functions from spherical deconvolution [12] were used for tractography. PVE maps were used in the tracking process to provide a better tracking domain as opposed to fractional anisotropy (FA)-based mask where streamline propagation is often prematurely halted in crossing regions.

Probabilistic Particle Filtering Tractography [9] was done seeding from the WM and GM interface (1 × 1 × 1 mm³, 2M seeds). The particle filtering tractography algorithm ensured that streamlines did not terminate prematurely in the WM by the application of a back-tracking rule to allow the tractography algorithm to find alternative pathways. Freesurfer [7] was used to parcellate the brain into 163 labels [5]. Subcortical regions were included to ensure an accurate representation of WM connections throughout the brain (e.g. thalamocortical radiations). The same reasoning was applied to the brain stem and cerebellum regions to ensure the inclusion of the corticospinal/corticocerebellar tracts within the graph. A 3 mm dilation was used to ensure a robust overlap between streamlines end-points (e.g. GM/WM interface) and anatomical labels [14]. Finally, streamlines and brain labels were loaded in FiberNavigator[1] [3].

3   Visualization

First, an iso-surface is derived from the T1-weighted image for contextual reference as shown in Fig. 1. Next, a spherical node (red) is positioned at the

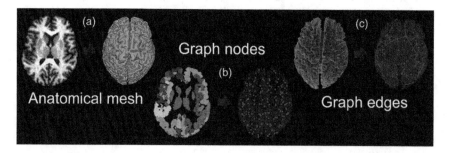

Fig. 1. Graph construction. (a) Mesh derived from anatomical T1 image. (b) Nodes derived from anatomical labels. (c) Edges derived from tractogram. (Color figure online)

---

[1] Open source software available at: chamberm.github.io/fibernavigator_single.html.

barycenter of each label. A default weighted connectivity matrix $\mathbf{M}$ is built by normalizing the number of streamlines linking each anatomical region[2]. A transfer function is responsible for mapping values of $\mathbf{M}$ towards edge thickness and opacity. The default view also re-sizes each node by its degree and a side panel shows a set of global graph metrics (e.g., mean degree, global efficiency).

Selecting a node instantly initiates the computation of node-related metrics (e.g. degree, strength, centrality, efficiency). In addition, selecting any 2 nodes immediately reveals the underlying streamlines forming the edge between them. An interactive global threshold (acting on the weights of $\mathbf{M}$) is also available, which automatically updates the global and local metrics of the network on the fly, as well as the visualization of the graph. Finally, to reduce visual ambiguity in node selection, nodes are depth sorted and color-graded in real-time according to the current viewpoint. Importantly, although very fast, the new framework is implemented on CPU using C++ and GLSL shaders, can run on a single core computer, and does not require any specific hardware. Experiments were performed on a laptop with the following specifications: System: Windows 8, Video card: Geforce GT 640 M memory 2 GB, NVIDIA Driver: 306.97, CPU: Intel(R)Core(TM) i7-3632QM @ 2,20 GHz, 16 GB RAM.

4   Results

Underlying streamlines linking 2 nodes are illustrated in Fig. 2. From left to right: corpus callosum (CC), optic radiation (OR) and corticospinal tract (CST). Controversial streamlines forming thick edges in the graph (number of streamlines in this case) are easily identified (e.g. Frontal Aslant Tract (FAT) [2]) and can potentially be removed from the network (e.g. $\mathbf{M}_{ij} = 0$) as shown in Fig. 3.

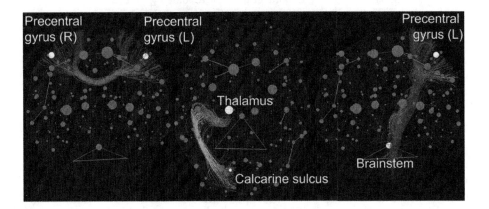

Fig. 2. Bundle selection using node picking (white).

---

[2] Demo available online at: www.youtube.com/watch?v=eZ2JubD25NA.

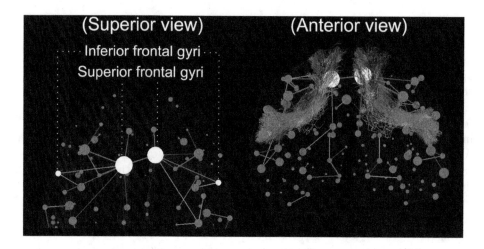

Fig. 3. Frontal aslant tract (FAT) [2] rapidly identified by the selection of 2 nodes.

Figure 4 shows two versions of the whole-brain network (i.e. unthresholded vs thresholded) as well as its associated global and nodal graph metrics (Tables 1 and 2). Given a specific node of interest (e.g. pre-central gyrus, Fig. 4 yellow), the user can instantaneously observe variations in the different metrics related to that node by dragging the threshold slider (2% threshold). A 30 frame-per-second (FPS) ratio was maintained during the process.

Table 1. Real-time global graph metrics

Metrics	Default graph	Thresholded graph (2%)
# of nodes	161	160
# of edges	4632	938
Density	0.36	0.07
Mean degree	62.59	12.68
Global efficiency	0.446	0.104

Finally, Fig. 5 shows how depth-sorting can help differentiate occipital nodes from frontal nodes. For any viewpoint, a transfer function assigns a color grading to each node based on their Z position in the scene. In this example, nodes located in the posterior aspect of the brain appear brighter than the ones located in the frontal lobe since the camera is looking at the brain from behind.

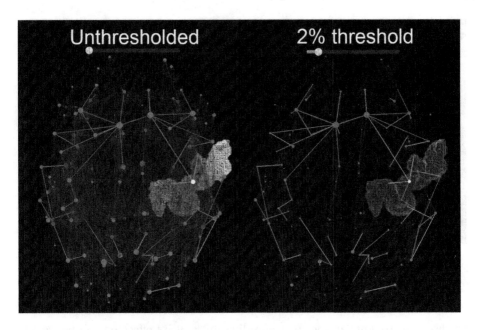

Fig. 4. Threshold graph visualization. Node sizes are recomputed on the fly according to their new strength. Yellow: pre-central gyrus (R). (Color figure online)

Table 2. Real-time local graph metrics (right pre-central gyrus)

Metrics	Default graph	Thresholded graph (2%)
Degree	88	24
Strength	3.73	3.49
Eigen centrality	0.165	0.161
Closeness centrality	0.919	0.691
Local efficiency	0.896	0.955

5   Discussion

To the best of our knowledge, this is the first visualization platform supporting comprehensive exploration of structural connectomics in real-time. The tool allows the user to easily prune undesired edges of the graph (e.g. false-positive streamlines). The mean FPS ratio was above 30 during all steps, indicating no latency. Initial piloting of the tool (by users new to graph theory) revealed the following consensus: hubs and underlying streamlines were easily identifiable by all. Moreover, participants were mostly curious how simple threshold manipulation altered local and global network metrics.

After discussing with neurosurgeons, the framework also incorporates various representation of $\mathbf{M}$ by allowing direct manipulation of bundle-specific edge

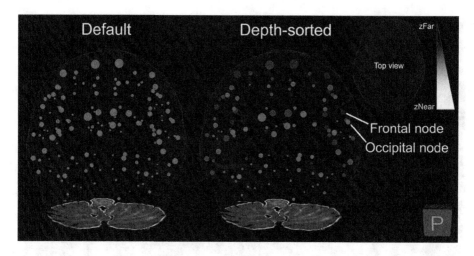

Fig. 5. Depth-sorted nodes provide increase visual cues when compared to default rendering.

weights (e.g. to simulate de- or re-myelination and its effect on the network). The current version also allows users to input a more general connectivity matrix (e.g. derived from other software or image modalities such as functional MRI or MEG). In other words, the users are not bound to a specific tractography pipeline to generate the aforementioned connectivity matrix. Moreover, it is important to specify that any set of brain parcellation can be used here (i.e. varying number of labels).

6   Conclusion

With the large variety of metrics and parameters involved in connectomics (e.g. weights of $\mathbf{M}$, threshold techniques [6]), the proposed growing visualization framework will also serve as a quality assurance tool for close inspection of data prior to launching massive analyses. From a clinical perspective, the proposed platform will also provide neurosurgeons with a better understanding of the effect of transecting pathways underlying critical hubs, and perhaps physiotherapists insight into the impact of strengthening a given edge on network characteristics.

References

1. Bullmore, E., Sporns, O.: Complex brain networks: graph theoretical analysis of structural and functional systems. Nat. Rev. Neurosci. **10**(3), 186 (2009)
2. Catani, M., Mesulam, M.M., Jakobsen, E., Malik, F., Martersteck, A., Wieneke, C., Thompson, C.K., et al.: A novel frontal pathway underlies verbal fluency in primary progressive aphasia. Brain **136**(8), 2619–2628 (2013)

3. Chamberland, M., Bernier, M., Fortin, D., Whittingstall, K., Descoteaux, M.: 3D interactive tractography-informed resting-state fMRI connectivity. Front. Neurosci. **9**, 275 (2015)
4. Côté, M.-A., Girard, G., Bor'e, A., Garyfallidis, E., Houde, J.-C., Descoteaux, M.: Tractometer: towards validation of tractography pipelines. Med. Image Anal. **17**(7), 844–857 (2013)
5. Destrieux, C., Fischl, B., Dale, A.M., Halgren, E.: A sulcal depth-based anatomical parcellation of the cerebral cortex. NeuroImage **47**, S151 (2009)
6. Drakesmith, M., Caeyenberghs, K., Dutt, A., Lewis, G., David, A.S., Jones, D.K.: Overcoming the effects of false positives and threshold bias in graph theoretical analyses of neuroimaging data. NeuroImage **118**, 313–333 (2015)
7. Fischl, B., Van Der Kouwe, A., Destrieux, C., Halgren, E., Ségonne, F., Salat, D.H., Evelina, B., et al.: Automatically parcellating the human cerebral cortex. Cereb. Cortex **14**(1), 11–22 (2004)
8. Fornito, A., Andrew, Z., Breakspear, M.: Graph analysis of the human connectome: promise, progress, and pitfalls. Neuroimage **80**, 426–444 (2013)
9. Girard, G., Kevin, W., Deriche, R.: Towards quantitative connectivity analysis: reducing tractography biases. Neuroimage **98**, 266–278 (2014)
10. Maier-Hein K., Neher P., Houde J-C., Cote M-A., Garyfallidis E., Zhong J., Chamberland M. et al. Tractography-based connectomes are dominated by false-positive connections. bioRxiv 084137 (2016)
11. Margulies, D.S., Böttger, J., Watanabe, A., Gorgolewski, K.J.: Visualizing the human connectome. NeuroImage **80**, 445–461 (2013)
12. Tournier, J.-D., Fernando, C., Connelly, A.: Robust determination of the fibre orientation distribution in diffusion MRI: non-negativity constrained super-resolved spherical deconvolution. NeuroImage **35**(4), 1459–1472 (2007)
13. Van Wijk, B.C.M., Cornelis, J.S.: Comparing brain networks of different size and connectivity density using graph theory. PloS One **5**(10), e13701 (2010)
14. Yeh, C.-H., Smith, R., Dhollander, T., Calamante, F., Connelly, A.: The influence of node assignment strategies and track termination criteria on diffusion MRI-based structural connectomics. In: International Symposium on Magnetic Resonance in Medicine (ISMRM), no. 0118 (2016)

Pairing-based Ensemble Classifier Learning using Convolutional Brain Multiplexes and Multi-view Brain Networks for Early Dementia Diagnosis

Anna Lisowska, Islem Rekik[✉],
and The Alzheimers Disease Neuroimaging Initiative

BASIRA Lab, CVIP Group, School of Science and Engineering, Computing,
University of Dundee, Dundee, UK
irekik@dundee.ac.uk
http://www.basira-lab.com

Abstract. The majority of works using brain connectomics for dementia diagnosis heavily relied on using structural (diffusion MRI) and functional brain connectivity (functional MRI). However, how early dementia affects the morphology of the cortical surface remains poorly understood. In this paper, we first introduce *multi-view morphological brain network* architecture which stacks multiple networks, each quantifying a cortical attribute (e.g., thickness). Second, to model the relationship between brain views, we propose a subject-specific *convolutional brain multiplex* composed of intra-layers (brain views) and inter-layers between them by convolving two consecutive views. By reordering the intra-layers, we generate different multiplexes for each subject. Third, to distinguish demented brains from healthy ones, we propose a *pairing-based ensemble classifier learning strategy*, which projects each pair of brain multiplex sets onto a low-dimensional space where they are fused, then classified. Our framework achieved the best classification results for the right hemisphere 90.8% and the left hemisphere 89.5%.

1 Introduction

Early diagnosis of brain dementia, specifically mild cognitive impairment (MCI) which may convert to Alzheimer's disease (AD), is critical for the early intervention, to prevent the onset of AD. Machine learning approaches have been successfully employed in diagnosing AD based on images obtained from MRI [1], which provide an efficient and non-invasive way for investigating neurological disorders at a whole-brain level. On a brain *connectional* level, network analysis of functional and structural brain connectivity (obtained from functional MRI (fMRI) and diffusion-weighted MRI (dMRI)) helped identify dementia biomarkers and brain connections affected by this neurodegenerative disorder [2]. Recently, more research has focused on accurate detection of early mild cognitive impairment (eMCI), which is essential for slowing down potential conversion to AD. For instance, [3] investigated the predictive power of various combinations of connectomic features, such as pairwise connectivity and maximum flow

© Springer International Publishing AG 2017
G. Wu et al. (Eds.): CNI 2017, LNCS 10511, pp. 42–50, 2017.
DOI: 10.1007/978-3-319-67159-8_6

between two brain regions, extracted from dMRI images for eMCI and normal control (NC) classification problem. On the other hand, [4] computed sparse temporal networks using sliding-window approach over a time series of resting-state functional MRI. [5] extended this work by additionally considering the high-order correlation between different pairs of brain regions. By combining low-order with high-order correlations, they further improved the classification accuracy of eMCI/NC.

Although dementia has been shown to affect neuronal connections in the brain as well as the cortical surface causing cortical thinning [6], research exploring *morphological connectivity* of the cortex is almost absent [1]. More specifically, how the shape of a cortical brain region gets affected *in relation* to the shape of another cortical brain region using various shape measurements (e.g., curvature, sulcal depth) remains somewhat unexplored. To address these limitations, we propose to use morphological cortical networks for dementia onset identification. Additionally to using one-layer network (considering only one morphological view, such as cortical thickness), we construct a multi-layer network (multiplex), consisting of multiple morphological views. Previous research showed that using multi-layer networks (i.e., stacking different networks) improved the prediction accuracy for disease identification when compared to using single view networks. Some of these works included classification of NC/MCI/AD using combination of features from MRI, PET, and CSF [7], structural inter- and intra-subject brain similarities in MRI [8], both confirming that multiplex network features yield better classification results in comparison to using low-level features. Other works, not concerned with MCI/AD, used multiplexes for simultaneous analysis of anatomical and functional brain networks [9] and varied frequency in fMRI to find important functional brain regions affected by schizophrenia [10].

However, none of these multiplex-based methods explored the *relationship* between two consecutive layers in the multiplex or *cortical morphology*. Specifically, to the best of our knowledge, no previous methods explored the similarity between layers in a typical multi-layer network for modeling brain connectivity [1]. We note that simple concatenation of multiple networks hinders the investigation of potentially *complex* changes in cortical regions, which might vary jointly or independently across different brain views as they become affected by dementia onset. Hence, we introduce *inter-layers* into a multiplex structure to capture the relationship between different brain views. Basically, each brain multiplex consists of different morphological views (intra-layers) and inter-layers splipped between two consecutive intra-layers, thereby quantifying the relationship between two consecutive brain views.

Since each multiplex is not invariant to the ordering of the intra-layers, we generate multiple multiplexes for each subject while considering all possible combinations of intra-layers, thereby capturing all relationship between different brain views in a more holistic manner. Fusing information from different brain multiplexes is crucial for more accurate identification of the demented brain state since each brain multiplex captures a unique relationship between brain views, which can help unravel the complex nature of brain disorders for more accurate

Table 1. Major mathematical notations used in this paper.

Mathematical notation	Definition
$\mathbf{V}$	Brain network (single view) in $\mathbb{R}^{n \times n}$
$\mathcal{M}$	Brain multiplex composed of intra-layers and convolutional inter-layers
$\mathbf{C}_{i,j}$	Convolutional intra-layer between consecutive brain network views $\mathbf{V}_i$ and $\mathbf{V}_j$ in $\mathcal{M}$
$\mathbb{M} = \{\mathcal{M}_1, \ldots, \mathcal{M}_N\}$	Subject-specific brain multiplexes with different orderings of intra-layers
$\mathbf{M}_k$	matrix in $\mathbb{R}^{d \times N_s}$ Containing the d multiplex features for all N_s training samples from multiplex $\mathcal{M}_k \in \mathbb{M}$
$\mathbf{M}_{k,l} = [\mathbf{M}_k, \mathbf{M}_l]$	Paired multiplex feature matrices derived from two training multiplexes in $\mathbb{M}$
$[\mathbf{B}_k, \mathbf{B}_l]$	CCA basis matrices spanning the canonical space where $\mathbf{M}_k$ and $\mathbf{M}_l$ are projected
$\mathbf{\Sigma}_{\mathbf{k},\mathbf{l}}$	Covariance matrix of paired training multiplex matrices $\mathbf{M}_k$ and $\mathbf{M}_l$
$\mathbf{W}_k$	Transformation matrix from the original multiplex space to the low-dimensional canonical multiplex space
$\mathbf{\Lambda}^2$	Diagonal matrix of eigenvalues (i.e., canonical correlations squared)
$\hat{\mathbf{M}}_k$	Canonical representation of multiplex $\mathcal{M}_k$ projected onto CCA space
$\hat{\mathbf{M}}_{k,l}$	Fused CCA-mapped multiplex feature matrices of original multiplexes $\mathcal{M}_k$ and $\mathcal{M}_l$
$\mathbf{I}$	Identity matrix in $\mathbb{R}^{d \times d}$

diagnosis. However, most existing network fusion methods often extract features independently from each network, and then simply concatenate them into a long feature vector for classification [1], while overlooking the correlation between them. To address this issue, we propose to use canonical correlation analysis (CCA) to map two sets of features into a shared space where they become more comparable [11,12]. CCA was shown to yield more discriminative features than any of the input feature vectors alone or their simple concatenation [12]. Since we are not restricted to only two sets of features as in [12], we propose a novel pairing-based CCA mapping of multiple sets of brain multiplexes, where each pair of multiplex sets is mapped onto a CCA space then fused. Ultimately, in the spirit of ensemble classifier learning, we input the fused multiplex features to train a linear classifier in each spanned CCA space.

Overall, we propose three fundamental contributions to the state-of-the-art of brain network analysis in order to identify dementia in its early stage: (1) brain multiplex structure based on cortical morphology, (2) pairing-based ensemble classifier learning strategy using CCA-mapped sets of brain connectomic features, and (3) giving new insights into how the early stage of MCI affects *morphological* brain connectivity in left and right cortical hemispheres.

2   Ensemble Classifier Using Paired CCA-Mapped Convolutional Brain Mutliplexes for eMCI/NC Classification

In this section, we introduce the concept of a convolutional brain multiplex and present our novel canonical ensemble classifier learning technique using paired sets of brain multiplexes. Matrices are denoted by boldface capital letters,

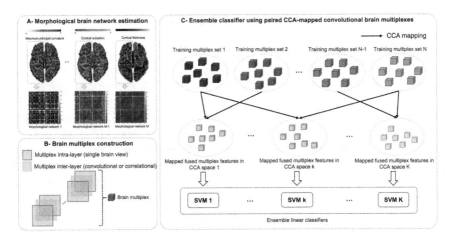

Fig. 1. *Pipeline of the proposed pairing-based ensemble classifier learning using fused convolutional brain multiplexes.* (A) Morphological brain network construction using different cortical attributes. (B) Brain multiplex construction. (C) We use canonical correlation analysis (CCA) to first project a pair of multiplex sets onto a common space where they become more comparable, then fuse them together to train a linear SVM classifier.

e.g., $\mathbf{X}$, and scalars are denoted by lowercase letters, e.g., x. We denote the transpose operator and the trace operator as $\mathbf{X}^T$ and $tr(\mathbf{X})$, respectively. For easy reference and enhancing the readability, we have summarized the major mathematical notations in Table 1. We illustrate in Fig. 1 the proposed framework for convolutional brain multiplex construction and pairing-based ensemble classifier learning using CCA mapping of sets of brain multiplexes.

• *Step 1: Convolutional brain multiplex construction and feature extraction.* In a generic way, we define a brain multiplex $\mathcal{M}$ using a set of M intra-layers $\{\mathbf{V}_1, \ldots, \mathbf{V}_M\}$, each representing a single view of the brain morphology, (i.e., cortical attribute), where between two consecutive intra-layers $\mathbf{V}_i$ and $\mathbf{V}_j$ we slide an inter-layer $\mathbf{C}_{i,j}$. This yields to the following multiplex architecture: $\mathcal{M} = \{\mathbf{V}_1, C_{1,2}, \mathbf{V}_2, \ldots, \mathbf{V}_i, \mathbf{C}_{i,j}, \mathbf{V}_j, \ldots, \mathbf{V}_M\}$. Each inter-layer is defined by convolving two consecutive intra-layers. Each element in row a and column b within the convolutional inter-layer matrix $\mathbf{C}_{i,j}$ between views $\mathbf{V}_i$ and $\mathbf{V}_j$ is defined as: $\mathbf{C}_{i,j}(a,b) = \sum_p \sum_q \mathbf{V}_i(p,q)\mathbf{V}_j(a-p+1, b-q+1)$. We note that for a specific multiplex, we are only allowed to explore similarities between consecutive layers. Hence, to explore the inter-relationship between all possible combinations of intra-layers, we generate for each subject N multiplexes through simply reordering the intra-layer networks, thereby generating an *ensemble multiplexes* $\mathbb{M} = \{\mathcal{M}_1, \ldots, \mathcal{M}_N\}$ (Fig. 1-B). Each subject-specific brain multiplex in $\mathbb{M}$ captures unique similarities between different morphological brain network views (e.g., sulcal depth network and cortical thickness network) that may not be present in a different multiplex. This will allow not only to explore how different brain views get altered by a specific disorder, but how their *relationship* might get affected.

Since the morphological brain connectivity matrices are symmetric (Fig. 1-A), we extract features from each multiplex by directly concatenating the weights of all connectivities in each triangular matrix. For each network of size $n \times n$, we extract a feature vector of size $(n \times (n-1)/2)$.

• **Step 2: Pairing-based ensemble classifier learning using canonical mapping of brain multiplex sets.** Since each multiplex $\mathcal{M}_k \in \mathbb{M}$ captures a unique and complex relationship between different brain network views, one needs to examine all morphological brain multiplexes in the ensemble $\mathbb{M}$. This will provide us with a more holistic understanding of how explicit morphological brain connections can be altered by dementia onset as well as how their implicit high-order (a connection of connections) relationship can be affected. However, due to complex nature of the multiplex structure, feature reduction method is required to reduce the redundancy of the data by extracting the most representative features. Instead of extracting features from different multiplexes independently, and motivated by the fact that canonical correlation analysis is efficient in analyzing and fusing associations between two sets of variables [11,12], we propose a pairing-based CCA mapping strategy of sets of multiplexes of our training samples for brain multiplex feature fusion.

Suppose that $\mathbf{M}_k \in \mathbb{R}^{d \times N_s}$ and $\mathbf{M}_l \in \mathbb{R}^{d \times N_s}$ are two training multiplex feature matrices derived from two different multiplexes in $\mathbb{M}$, where N_s denotes the number of training samples. For each pair of multiplexes $\mathbf{M}_{k,l} = [\mathbf{M}_k, \mathbf{M}_l]$, we define their covariance matrix $\mathbf{\Sigma}_{k,l} = \begin{pmatrix} cov(\mathbf{M}_k) & cov(\mathbf{M}_k, \mathbf{M}_l) \\ cov(\mathbf{M}_l, \mathbf{M}_k) & cov(\mathbf{M}_l) \end{pmatrix}$, where $cov(\mathbf{M}_k) = \mathbf{M}_k \mathbf{M}_k^T$ denotes the within-set covariance matrix of $\mathbf{M}_k$, and $cov(\mathbf{M}_k, \mathbf{M}_l) = \mathbf{M}_k \mathbf{M}_l^T$ denotes the between-set covariance matrix of $\mathbf{M}_k$ and $\mathbf{M}_l$. To map both training multiplex matrices onto a space where the respective distributions of their features are more 'aligned' and easily comparable, we aim to maximize the pair-wise correlation across the two matrices $\mathbf{M}_k$ and $\mathbf{M}_l$: $corr(\hat{\mathbf{M}}_k, \hat{\mathbf{M}}_l) = \frac{cov(\hat{\mathbf{M}}_k, \hat{\mathbf{M}}_l)}{var(\hat{\mathbf{M}}_k) \cdot var(\hat{\mathbf{M}}_l)}$, where $\hat{\mathbf{M}}_k$ denotes the linear CCA mapping of the multiplex feature matrix $\mathbf{M}_k$ to the canonical shared space using the estimated transformation matrix $\mathbf{W}_k^T$ such that $\hat{\mathbf{M}}_k = \mathbf{W}_k^T \mathbf{M}_k$. Similarly, the second set of training multiplex features $\mathbf{M}_l$ is mapped using the estimated transformation matrix $\mathbf{W}_l^T$. More precisely, $cov(\hat{\mathbf{M}}_k, \hat{\mathbf{M}}_l)$ is defined as $\mathbf{W}_k^T cov(\mathbf{M}_k, \mathbf{M}_l) \mathbf{W}_l$, $var(\hat{\mathbf{M}}_k)$ as $\mathbf{W}_k^T cov(\mathbf{M}_k) \mathbf{W}_k$, and $var(\hat{\mathbf{M}}_l)$ as $\mathbf{W}_l^T cov(\mathbf{M}_l) \mathbf{W}_l$.

Both canonical transformation matrices are estimated through maximizing the covariance between the mapped multiplex feature matrices $\hat{\mathbf{M}}_k$ and $\hat{\mathbf{M}}_l$, constrained to $var(\hat{\mathbf{M}}_l) = var(\hat{\mathbf{M}}_k) = I$, using Lagrange multipliers. This is achieved through solving the following eigenvector equations:

$$\begin{cases} cov(\mathbf{M}_k)^{-1} cov(\mathbf{M}_k, \mathbf{M}_l) cov(\mathbf{M}_l)^{-1} cov(\mathbf{M}_l, \mathbf{M}_k) \hat{\mathbf{W}}_k = \mathbf{\Lambda}^2 \hat{\mathbf{W}}_k \\ cov(\mathbf{M}_l)^{-1} cov(\mathbf{M}_l, \mathbf{M}_k) cov(\mathbf{M}_k)^{-1} cov(\mathbf{M}_k, \mathbf{M}_l) \hat{\mathbf{W}}_l = \mathbf{\Lambda}^2 \hat{\mathbf{W}}_l \end{cases},$$

where $\hat{\mathbf{W}}_k$ and $\hat{\mathbf{W}}_l$ denote the eigenvectors and $\mathbf{\Lambda}^2$ represent the diagonal matrix of eigenvalues (i.e., canonical correlations squared). The dimension of the canonical shared space is defined as the rank of covariance matrix between both multiplex feature matrices. Ultimately, each transformation matrix $\mathbf{W}_k$ is generated through sorting the eigenvectors in $\hat{\mathbf{W}}_k$ with non-zero eigenvalues. To perform paired multiplex feature fusion in the canonical space, we simply concatenate the transformed multiplex features as follows:

$$\hat{\mathbf{M}}_{k,l} = \begin{pmatrix} \hat{\mathbf{M}}_k \\ \hat{\mathbf{M}}_l \end{pmatrix} = \begin{pmatrix} \mathbf{W}_k^T \mathbf{M}_k \\ \mathbf{W}_l^T \mathbf{M}_l \end{pmatrix} = \begin{pmatrix} \mathbf{W}_k & 0 \\ 0 & \mathbf{W}_l \end{pmatrix}^T \begin{pmatrix} \mathbf{M}_k \\ \mathbf{M}_l \end{pmatrix}$$

Next, we use each fused pair of training multiplex feature matrices $\hat{\mathbf{M}}_{k,l}$ to train a linear support vector machine (SVM) classifier (Fig. 1). Noting that for each training subject we have N multiplexes estimated, we perform C_N^2 mappings of each pair of multiplexes in $\mathbb{M}$. Subsequently, a linear SVM classifier is learned for each pair of multiplexes. In the testing stage, we use the learned canonical transformation matrices to respectively map each pair of testing multiplex feature vector onto their corresponding CCA space where they are fused and then communicated to an SVM classifier. Finally, we average all soft scores by ensemble SVM classifiers to determine the label of the testing subject.

3 Results and Discussion

Evaluation dataset. We used leave-one-out cross validation to evaluate the proposed classification framework on 76 subjects (35 eMCI and 41 NC) from ADNI GO public dataset[1], each with structural T1-w MR image [13]. We note that the 35 eMCI samples comprise the first and last acquisition timepoints for 18 different eMCI subjects, which are largely spaced out in time. Hence, we assume that these two distant timepoints can simulate two different eMCI subjects. We used FREESURFER to reconstruct both right and left cortical hemispheres for each subject from T1-w MRI. Then we parcellated each cortical hemisphere into 35 cortical regions using Desikan-Killiany Atlas. We defined $N = 6$ multiplexes, each using $M = 4$ cortical network views. For each cortical attribute (signal on the cortical surface), we compute the strength of the morphological network connection linking i^{th} ROI to the jth ROI as the absolute difference between the averaged attribute values in both ROIs. Multiplex $\mathcal{M}_1$ includes cortical attribute views $\{\mathbf{V}_1, \mathbf{V}_2, \mathbf{V}_3, \mathbf{V}_4\}$, $\mathcal{M}_2$ includes $\{\mathbf{V}_1, \mathbf{V}_2, \mathbf{V}_4, \mathbf{V}_3\}$, $\mathcal{M}_3$ includes $\{\mathbf{V}_1, \mathbf{V}_3, \mathbf{V}_4, \mathbf{V}_2\}$, $\mathcal{M}_4$ includes $\{\mathbf{V}_1, \mathbf{V}_3, \mathbf{V}_2, \mathbf{V}_4\}$, $\mathcal{M}_5$ includes $\{\mathbf{V}_1, \mathbf{V}_4, \mathbf{V}_2, \mathbf{V}_3\}$, and $\mathcal{M}_6$ includes $\{\mathbf{V}_1, \mathbf{V}_4, \mathbf{V}_3, \mathbf{V}_2\}$. For each cortical region, $\mathbf{V}_1$ denotes the maximum principal curvature brain view, $\mathbf{V}_2$ denotes the mean cortical thickness brain view, $\mathbf{V}_3$ denotes the mean sulcal depth brain view, and $\mathbf{V}_4$ denotes the mean of average curvature.

Comparison methods and evaluation. For our eMCI/NC classification task, we benchmarked our pairing-based ensemble classifier strategy against: (1) using single SVM trained on each brain view, and on the concatenated views, (2)

---

[1] http://adni.loni.usc.edu.

Table 2. eMCI/NC classification accuracy using our method and different comparison methods

Classifier	Method	Left Hemisphere				Right Hemisphere			
		Accuracy (%)	AUC	Sensitivity (%)	Specificity (%)	Accuracy (%)	AUC	Sensitivity (%)	Specificity (%)
Single SVM	View 1	68.4	75.9	65.7	70.7	75.0	85.1	82.9	68.3
	View 2	77.6	83.9	82.9	73.2	81.6	85.3	85.7	78.0
	View 3	77.6	83.5	77.1	78.0	71.1	87.2	74.3	68.3
	View 4	53.9	53.5	0.0	100.0	53.9	46.3	0.0	100.0
	All Views Concatenated	81.6	88.3	85.7	78.0	86.8	89.1	88.6	85.4
Ensemble classifiers	Views	81.6	90.5	65.7	**95.1**	82.9	94.1	74.3	**90.2**
	Correlation	81.6	88.1	80.0	82.9	85.5	**96.7**	**91.4**	80.5
	Convolution	85.5	91.1	82.9	87.8	76.3	86.4	77.1	75.6
Ensemble paired classifiers	Views	86.8	90.7	88.6	85.4	85.5	93.2	85.7	85.4
	Correlation	78.9	87.0	80.0	78.0	84.2	96.3	88.6	80.5
	Convolution	80.3	88.6	80.0	80.5	72.4	82.0	74.3	70.7
Ensemble CCA paired classifiers	Views	80.3	86.9	77.1	82.9	**90.8**	95.7	**91.4**	**90.2**
	Correlation	85.5	89.7	88.6	82.9	84.2	94.8	88.6	80.5
	Convolution	**89.5**	**92.2**	**91.4**	87.8	84.2	90.0	85.7	82.9

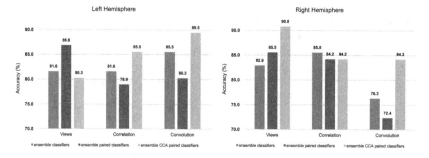

Fig. 2. *Classification accuracies for our proposed pairing-based ensemble classifier learning of CCA-mapped brain features and comparison ensemble classifier methods.* Views: concatenated brain views. Correlation: correlational brain multiplexes. Convolution: Convolutional brain multiplexes. Ensemble classifiers: one SVM trained for each view (or multiplex) without any pairing strategy or CCA mapping. Ensemble paired classifiers: pairing different views (or multiplexes) without CCA mapping. Ensemble CCA paired classifiers: pairing different views (or multiplexes) with CCA mapping.

ensemble SVM classifiers (without the pairing or CCA mapping strategies), and (3) ensemble paired SVM classifiers (without CCA mapping). For each of these methods, we generate three classification results using: (1) features from brain views, (2) features from correlational multiplexes (inter-layer computed using Pearson correlation), and (3) features from convolutional multiplexes (inter-layer computed using 2D convolution). For evaluation, we report in Table 2 the prediction accuracy, the area under the receiver operating characteristic (ROC) curve, the sensitivity and specificity of the eMCI/NC classification task. In Fig. 2, we specifically show the comparison of classification accuracy for ensemble, pairing-based ensemble and pairing-based ensemble with CCA-mapping classification based on views, correlational multiplexes and convolutional multiplexes. Our proposed ensemble CCA paired classifiers framework outperformed all comparison methods (89.5%) when using morphological multiplexes of the left hemisphere, while the best performance (90.8%) was achieved using multi-view brain network (i.e., stacked brain views) of the right hemisphere. These results may indicate that there is a difference in the level of complexity of the disease progression across both hemispheres, i.e., different morphological properties of the brain change in more consistent manner across the different views in the case of right hemisphere, while these interactions are more complex in the early progression of dementia in the left hemisphere. We also show that using the combination of different morphological views resulted in better classification accuracies in case of all reported methods than those based on single brain views.

4   Conclusion

We propose a novel pairing-based ensemble classifier strategy that fuses morphological multi-view brain networks as well as convolutional brain multiplexes for distinguishing between eMCI patients and healthy controls. The performance of our method gave us insights into how dementia might affect the right and the left hemispheres in its early stage: complex connectional alterations in cortical morphology spanning multiple cortical attributes of the left hemisphere (captured by the multiplex), and simple alterations across different brain views in the right hemisphere (captured by the morphological multi-view network). In our future work, we will integrate functional and diffusion networks in our multiplex structure to explore how the relationship between *multimodal* connectomic views is altered with dementia onset.

References

1. Brown, C., Hamarneh, G.: Machine learning on human connectome data from MRI (2016). arxiv:1611.08699v1
2. Bullmore, E., Sporns, O.: Complex brain networks: graph theoretical analysis of structural and functional systems. Nat. Neurosci. **10**, 186–198 (2009)
3. Prasad, G., Joshi, S.H., Nir, T.M., Toga, A.W., Thompson, P.M.: Brain connectivity and novel network measures for Alzheimer's disease classification. Neurobiol. Aging **36**(Supplement 1), S121–S131 (2015). Novel Imaging Biomarkers for Alzheimer's Disease and Related Disorders (NIBAD)
4. Wee, C.Y., Yang, S., Yap, P.T., Shen, D.: Sparse temporally dynamic resting-state functional connectivity networks for early MCI identification. Brain Imag. Behav. **10**, 342–356 (2016)
5. Chen, X., Zhang, H., Gao, Y., Wee, C.Y., Li, G., Shen, D.: The Alzheimer's disease neuroimaging initiative: high-order resting-state functional connectivity network for MCI classification. Hum. Brain Mapp. **37**, 3282–3296 (2016)
6. Querbes, O., Aubry, F., Pariente, J., Lotterie, J., Demonet, J., Duret, V., Puel, M., Berry, I., Fort, J., Celsis, P.: The Alzheimer's disease neuroimaging initiative: early diagnosis of alzheimer's disease using cortical thickness: impact of cognitive reserve. Brain **132**, 2036 (2009)
7. Zippo, E.G., Castiglioni, I.: Integration of 18FDG-PET metabolic and functional connectomes in the early diagnosis and prognosis of the Alzheimer's disease. Current Alzheimer Res. **13**, 487–497 (2016)

8. La Rocca, M., et al.: A multiplex network model to characterize brain atrophy in structural MRI. In: Mantica, G., Stoop, R., Stramaglia, S. (eds.) Emergent Complexity from Nonlinearity, in Physics, Engineering and the Life Sciences. Springer Proceedings in Physics, vol. 191. Springer, Cham (2017)

9. Crofts, J.J., Forrester, M., O'Dea, R.D.: Structure-function clustering in multiplex brain networks. EPL (Europhysics Letters) **116**, 18003 (2016)

10. Domenico, M.D., Sasai, S., Arenas, A.: Mapping multiplex hubs in human functional brain networks. Front. Neurosci. **10**, 326 (2016)

11. Haghighat, M., Abdel-Mottaleb, M., Alhalabi, W.: Fully automatic face normalization and single sample face recognition in unconstrained environments. Expert Syst. Appl. **47**, 23–34 (2016)

12. Zhu, X., Suk, H.I., Lee, S.W., Shen, D.: Canonical feature selection for joint regression and multi-class identification in Alzheimer's disease diagnosis. Brain Imaging Behav. **10**, 818–828 (2016)

13. Mueller, S.G., Weiner, M.W., Thal, L.J., Petersen, R.C., Jack, C., Jagust, W., Trojanowski, J.Q., Toga, A.W., Beckett, L.: The Alzheimer's disease neuroimaging initiative. Neuroimaging Clin. N. Am. **10**, 869–877 (2005)

High-order Connectomic Manifold Learning for Autistic Brain State Identification

Mayssa Soussia[1,2] and Islem Rekik[1(✉)]

[1] BASIRA Lab, CVIP Group, School of Science and Engineering, Computing,
University of Dundee, Dundee, UK
`irekik@dundee.ac.uk`
[2] Department of Electrical Engineering,
The National Engineering School of Tunis (ENIT), Tunis, Tunisia
`http://www.basira-lab.com`

Abstract. Previous studies have identified *disordered* functional (from fMRI) and structural (from diffusion MRI) brain connectivities in Autism Spectrum Disorder (ASD). However, 'shape connections' between brain regions were rarely investigated in ASD – e.g., how morphological attributes of a specific brain region (e.g., sulcal depth) change in relation to morphological attributes in other regions. In this paper, we use conventional T1-w MRI to define morphological connectivity networks, each quantifying shape similarity between different cortical regions for a specific cortical attribute at both *low-order* and *high-order* levels. For ASD identification, we present a connectomic manifold learning framework, which learns multiple kernels to estimate a similarity measure between ASD and normal controls (NC) connectomic features, to perform dimensionality reduction for clustering ASD and NC subjects. We benchmark our ASD identification method against supervised and unsupervised state-of-the-art methods, while depicting the most discriminative high- and low-order relationships between morphological regions in the left and right hemispheres.

1 Introduction

Autism spectrum disease (ASD) is a neurodevelopmental disorder characterized by altered cognitive functions, specifically difficulties in learning and impairment in social interaction. The centers for Disease Control and Prevention (CDC) estimates autism's prevalence as 1 in 68 children in the United States. Recently, interest in understanding how ASD alters connectivity between different brain regions has grown with the development of important technological and methodological neuroimaging tools. Most of connectomic studies on ASD in the literature [1–3] have mainly focused on structural and functional connectivity (FC) derived from diffusion tensor imaging (DTI) and functional magnetic resonance imaging (fMRI), respectively. For example, [4] used functional MRI to quantify consistent spatial temporal FC patterns to distinguish between ASD subjects and normal controls (NC). Another work [5] applied a unified connectivity framework

© Springer International Publishing AG 2017
G. Wu et al. (Eds.): CNI 2017, LNCS 10511, pp. 51–59, 2017.
DOI: 10.1007/978-3-319-67159-8_7

on DTI that learns the underlying patterns of ASD pathology through projective non-negative decomposition into sets of discriminative, developmental and reconstructive components. However, one of the potential limitations of solely relying on fMRI or DTI are: (1) fMRI has low signal-to-noise ratio possibly induced by non-neural noise, hence its derived functional connectivity strength between pairs of ROIs can be spurious or noisy, and (2) fiber tractography methods can produce largely variable and somewhat biased structural brain networks [6]. As an alternative, we propose a different type of brain network: a *morphological* network *solely* constructed from T1-w MRI. Since recent research on ASD reported brain cortical thickness changes in autism during early childhood [7], we specifically propose to build different morphological networks based on the morphology of the cortical surface, where each network is associated with a unique low-order cortical attribute such as sulcal depth or cortical thickness. However, since simply concatenating morphological brain networks overlooks how their relationship might be affected at a *higher-order* level by autism, we introduce *morphological high-order brain networks* for autism identification. Unlike functional high-order networks which model the dynamic brain activity within a time-window [8], our high-order network (HON) is able to characterize more complex interaction patterns among brain regions *in morphology*.

On the other hand, a very limited number of works used machine-learning methods on human connectome data from MRI for ASD/NC classification [9], in a supervised manner. For instance, [10] used a functional network estimated from resting state fMRI to distinguish between ASD subjects and healthy controls. [11], which adopted a network regularized support vector machine, used DTI to identify faulty sub-networks associated with ASD. However, while the majority of supervised machine-learning techniques are somewhat limited in terms of scalability as they require reliable and accurate labeling of medical data, unsupervised learning techniques can provide decision support for early intervention, help develop data-driven guidelines for care plan management, and help group patients by similar non-semantic features (i.e., latent representation of brain disorder group or subgroup), to enable comparative effectiveness research (e.g., of medications) [12]. From a connectomic perspective, very few studies applied unsupervised learning methods for brain disease applications [9]. For instance, [13] computed spectral graph clustering to identify significant connectome modules for different brain disorder groups (Alzheimer's disease (AD) and Significant Memory Concern (SMC)). Another work [14] used a multi-view spectral clustering to group functional and structural brain networks of traumatic brain injury (TBI) patients. However, to the best of our knowledge, no previous unsupervised learning methods were used to distinguish between autistic and healthy brains [9].

To fill this gap, we propose a high-order morphological connectomic manifold learning for ASD identification using a novel unsupervised data clustering method called single-cell interpretation via multikernel learning (SIMLR) [15]. SIMLR has many appealing aspects. First, it inputs high-order networks and efficiently learns a similarity matrix between networks by combining multiple kernels which provides a better fit to the inherent statistical distribution of the HON data. Second, it is scalable and separates subpopulations more accurately than conventional

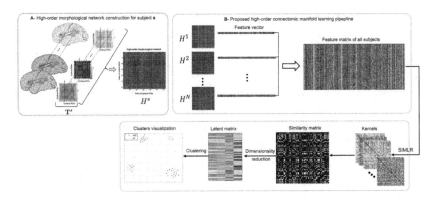

Fig. 1. *Illustration of the proposed high-order connectomic manifold learning for autistic brain state identification.* (A) High-order morphological network construction using multiple brain networks, each measuring a unique cortical attribute (e.g., thickness) on the cortical surface. These are stacked together to form a morphological brain tensor $\mathcal{T}^s$ for subject s. (B) Given the high-order feature matrix of all subjects, we used SIMLR [15] to learn proper weights for multiple kernels, which measure different distances between subjects. Next, we use the learned kernels to construct a symmetric similarity matrix $\mathbf{S}$ between subjects. SIMLR imposes a low-rank constraint on $\mathbf{S}$ such that different populations of the input data will be embedded into independent block-diagonal structure that clusters similar samples. This outputs a latent data representation in a low-dimensional space, which is inputted to a clustering algorithm. Each point in the 2D scatter plot represents an ASD or NC subject, and the corresponding colors represent the true labels in each cluster.

methods (e.g., PCA or t-SNE). Third, it improves weak similarities between samples through graph diffusion, which adds transitive similarities between dissimilar regions that have many similar neighboring regions. We compare our framework with both supervised and unsupervised disease identification techniques. To the best of our knowledge, this is the first work that: (1) defines high-order *morphological* brain networks, (2) jointly integrates multiple cortical morphological brain networks for autism identification, and (3) utilizes unsupervised SIMLR technique on ASD connectomic data.

2 High-Order Connectomic Manifold Learning for Unsupervised Clustering of Autistic and Healthy Brains

In this section, we present the high-order connectomic manifold learning for ASD identification using multiple kernels based on SIMLR technique introduced in [15]. We denote tensors by boldface Euler script letters, e.g., $\mathcal{X}$. Matrices are denoted by boldface capital letters, e.g., $\mathbf{X}$, and scalars are denoted by lowercase letters, e.g., x. For easy reference and enhancing the readability, we have summarized the major mathematical notations in Table 1. Figure 1 illustrates the proposed pipeline for ASD/NC identification in four major steps. (1) construction

Table 1. Major mathematical notations used in this paper

Mathematical notation	Definition
$\mathcal{T}^{\mathbf{s}}$	brain tensor of subject s in $\mathbb{R}^{n_r \times n_r \times n_v}$
$\mathbf{X}^k$	brain network in $\mathbb{R}^{n_r \times n_r}$ denoting the k-th frontal-view of tensor $\mathcal{T}$
$\mathbf{H}^s$	high-order morphological brain network for subject s
$\mathbf{h}_s$	high-order feature vector extracted from the upper triangular part of $\mathbf{H}^s$
$\mathbf{K}_l$	l-th learning kernel in $\mathbb{R}^{n \times n}$
n	number of subjects
$\mathbf{S}$	similarity matrix in $\mathbb{R}^{n \times n}$ for connectomic manifold learning
$\mathbf{L}$	latent matrix in $\mathbb{R}^{n \times c}$
c	number of clusters
m	number of kernels
$\mathbf{w}$	weighting vector of the kernels in $\mathbb{R}^m$
$\mathbf{I}_n$	identity matrix in $\mathbb{R}^{n \times n}$

of low-order morphological network (2) construction of high-order morphological network (3) feature extraction (4) connectomic manifold learning using SIMLR.

Low-order morphological network construction. For each subject s, we construct a brain tensor $\mathcal{T}^{\mathbf{s}}$ of size $\mathbb{R}^{n_r \times n_r \times n_v}$ for each cortical hemisphere, where n_r is the number of cortical regions of interest (ROIs) and n_v is the number of the tensor frontal views. Basically, for each cortical attribute (e.g., thickness), we construct a morphological brain network that constitutes a frontal view in $\mathcal{T}^{\mathbf{s}}$. Let x_i^k and x_j^k denote the mean of a cortical attribute of the i-th ROI and the j-th ROI in the k-th frontal view respectively. We then compute the absolute difference between x_i^k and x_j^k which depicts the connectivity strength between ROIs i and j. An element in the i-th row and j-th column of the k-th frontal view $\mathbf{X}^k$ is defined as: $\mathbf{X}_{ij}^k = |x_i^k - x_j^k|$.

High-order morphological network construction (HON). As the low-order network is unable to reveal the intrinsic similarities between more than a pair of ROIs, we propose to construct a high-order morphological network based on Pearson correlation to detect more complex interaction patterns between multiple brain regions. In addition to maintaining the pairwise relationship between ROIs in the same morphological view, the morphological HON underlines the relationship between ROIs across different views. Let y_{ij}^s denote the vector of the s-th subject corresponding to the connectivity strength between the i-th and j-th ROIs across all views. Each row in the high-order network $\mathbf{H}^s$ represents a pair of ROIs (i, j) and each column denotes a pair of ROIs (p, q). For a subject s, an element in H^s is defined using the Pearson's correlation coefficient as $\mathbf{H}_{ij,pq}^s = corr(y_{ij}^s, y_{pq}^s)$. We note that the entries $\mathbf{H}_{ij,pq}^s$ of the HON matrix indicate the influence of the connectivity strength between the i-th and j-th ROI on the connectivity strength between the p-th and q-th ROI. Thus, it underlines the higher order relationship between multiple ROIs.

Feature Extraction. For each subject, features are extracted in a naive way. Due to their symmetry, we concatenate the upper triangle elements of the HON matrix for subject s into a long feature vector $\mathbf{h}^s$. The weights on the diagonal are set to zero to avoid self-connectedness. Next, using K-fold data partition scheme, the extracted features of all ASD and NC subjects, while excluding the K-th fold, are fed into SIMLR.

High-order connectomic manifold learning. In this section, we briefly present the framework introduced in [15] and how we adapted it to our aim. The main idea of SIMLR is to learn a pairwise similarity matrix of size $n \times n$ from an input matrix of size $n \times d$ where n is the number of subjects and d is the dimension of their associated feature vectors. This allows to learn the connectomic manifold where all HON features $\{\mathbf{h}^1, \ldots, \mathbf{h}^n\}$ are nested. Instead of using one predefined distance metric which may fail to capture the nonlinear relationship in the data, we use multiple Gaussian kernels with learned weights to better explore in depth the similarity patterns among ASD and NC HONs. In other words, adopting multiple kernels allows to better fit the true underlying statistical distribution of the input matrix of high-order features. Additionally, constraints are imposed on kernel weights to avoid a single kernel selection [15]. The Gaussian kernel is expressed as follows: $\mathbf{K}(\mathbf{h}^i, \mathbf{h}^j) = \frac{1}{\epsilon_{ij}\sqrt{2\pi}} e^{\left(-\frac{|\mathbf{h}^i - \mathbf{h}^j|^2}{2\epsilon_{ij}^2}\right)}$, where $\mathbf{h}^i$ and $\mathbf{h}^j$ denote the feature vectors of the i-th and j-th subjects respectively and ϵ_{ij} is defined as: $\epsilon_{ij} = \sigma(\mu_i + \mu_j)/2$, where σ is a tuning parameter and $\mu_i = \frac{\sum_{l \in KNN(\mathbf{h}^i)} |\mathbf{h}^i - \mathbf{h}^j|}{k}$, where $KNN(\mathbf{h}^i)$ represents the top k neighboring subjects of subject i. The computed kernels are then averaged to further learn the similarity matrix $\mathbf{S}$ through an optimization framework formulated as follows:

$$\min_{\mathbf{S}, \mathbf{L}, \mathbf{w}} \sum_{i,j} -w_l \mathbf{K}_l(\mathbf{h}^i, \mathbf{h}^j) \mathbf{S}_{ij} + \beta ||\mathbf{S}||_F^2 + \gamma \mathbf{tr}(\mathbf{L}^T(\mathbf{I}_n - \mathbf{S})\mathbf{L}) + \rho \sum_l w_l \log w_l \quad (1)$$

Subject to: $\sum_l w_l = 1$, $w_l \geq 0$, $\mathbf{L}^T\mathbf{L} = \mathbf{I}_c$, $\sum_j \mathbf{S}_{ij} = 1$, and $\mathbf{S}_{ij} \geq 0$ for all (i, j), where:

1. $\sum_{i,j} -w_l \mathbf{K}_l(\mathbf{h}^i, \mathbf{h}^j) \mathbf{S}_{ij}$ refers to the relation between the similarity and the kernel distance with weights w_l between two subjects. The learned similarity should be small if the distance between a pair of subjects is large.
2. $\beta ||\mathbf{S}||_F^2$ denotes a regularization term that avoids over-fitting the model to the data.
3. $\gamma \mathbf{tr}(\mathbf{L}^T(\mathbf{I}_n - \mathbf{S})\mathbf{L})$: $\mathbf{L}$ is the latent matrix of size $n \times c$ where n is the number of subjects and c is the number of clusters. The matrix $(\mathbf{I}_n - \mathbf{S})$ denotes the graph Laplacian.
4. $\rho \sum_l w_l \log w_l$ imposes constraints on the kernel weights to avoid selection of a single kernel.

An alternating convex optimization is adopted where each variable is optimized while fixing the other variables until convergence [15]. Once, the similarity matrix $\mathbf{S}$ is obtained, a dimensionality reduction is performed on $\mathbf{S}$ using t-SNE [16]. In other words, the data is projected onto a lower dimension that preserves the similarity depicted in $\mathbf{S}$ resulting in an $n \times c$ latent matrix $\mathbf{L}$. For visualization, the same algorithm is used to create an embedding of $\mathbf{S}$ in a 2D space. A K-means clustering is then applied to the latent matrix $\mathbf{L}$ to cluster similar subjects and assess the concordance with the true labels (Fig. 1). It should be noted that the true labels were only used in the form of distinct colors to intuitively visualize the groups in (Fig. 1).

3   Results and Discussion

Evaluation dataset and method parameters. We evaluated the proposed clustering framework on 80 subjects (40 ASD and 40 NC) from Autism Brain Imaging Data Exchange (ABIDE I)[1] public dataset, each with structural T1-w MR image [17]. We used FREESURFER to reconstruct both right and left cortical hemispheres for each subject from T1-w MRI. Then we parcellated each cortical hemisphere into 35 cortical regions using Desikan-Killiany Atlas. For each subject, we generated $n_v = 4$ cortical morphological networks: $\mathbf{X}^1$ denotes the maximum principal curvature brain view, $\mathbf{X}^2$ denotes the mean cortical thickness brain view, $\mathbf{X}^3$ denotes the mean sulcal depth brain view, and $\mathbf{X}^4$ denotes the mean of average curvature. For SIMLR parameters, using a nested grid search, we set the number of clusters to $c = 4$. We used $m = 21$ kernels where each kernel is determined by a set of hyperparameters ($\sigma = 1 : 0.25 : 2.5$, number of top KNN neighbors in $\{10, 12, 14\}$), where σ is the variance parameter of the Gaussian function.

Method evaluation and comparison methods. To evaluate the reproducibility of our high-order connectomic manifold learning and clustering framework, we used two k-fold cross-validation schemes ($k = 5$ and $k = 10$) using randomized partitioning of data samples. The process was repeated 30 times and the average classification performance reported as final result for all comparison methods. We first compared our ASD/NC clustering method with Ward's linkage clustering [18], a widely used hierarchical clustering algorithm which optimizes a Euclidean objective function as a criterion for merging a pair of clusters at each step. This method was previously used for clustering functional HON networks for Alzheimer's disease diagnosis in [8]. Second, we compared the ASD/NC segregation efficiency of our method with two classification frameworks based on supervised linear support vector machine (SVM) classifier. Specifically, the first supervised method learns a single SVM using training connectomic features. To further evaluate SIMLR in a supervised manner, we propose a SIMLR-based ensemble classifier learning framework, where we use SIMLR to cluster the training data into different clusters, and then train an SVM classifier for each training

---

[1] http://fcon_1000.projects.nitrc.org/indi/abide/.

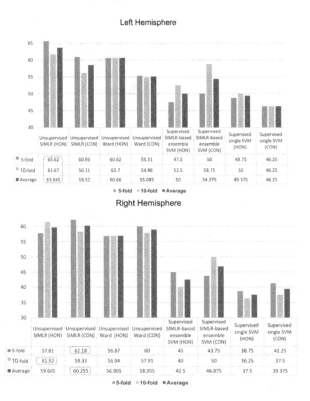

Fig. 2. *ASD identification accuracy using our method and comparison supervised and unsupervised methods.* We evaluated each of these methods on (i) the concatenated low-order morphological networks (i.e., 4 views) that we term with CON, and (ii) the high-order morphological networks (HON).

cluster. In the testing stage, we use label majority voting by all trained SVM classifiers to label an input testing subject. Each of these methods was evaluated on (i) the concatenated low-order morphological networks (i.e., 4 views) that we term with CON, and (ii) the high-order morphological networks (HON). Figure 2 displays ASD identification accuracies of all methods.

For the left hemisphere (LH), our method (Fig. 2–unsupervised SIMLR HON) had the best performance in distinguishing between ASD/NC subjects among all methods using both 5-fold and 10-fold cross-validation schemes, with an average performance of 63.64%. We note that the accuracies increased in average with HON across all methods, which might indicate that LH has more discriminative regions at a higher-order morphological level, except for the supervised SIMLR based ensemble SVM which scored better with CON. The low performance of supervised SIMLR-based ensemble SVM can be explained by the fact that SIMLR tends to produce more homogenous clusters, hence creating a non-balanced data samples for SVM training. This points to the imbalanced data issue for training supervised methods. On the other hand, results for the right hemisphere (RH) were better in average with unsupervised SIMLR CON. The performance of other methods also peaked when using CON features, except for supervised SIMLR ensemble based SVM. This might indicate that morphological changes due to ASD in RH regions occur at a low-order morphological connectivity level rather than a higher order level. In other words, the RH pairwise connectivity strength between regions in the same view depicts better the changes associated with autism than the high order relationship between regions of different views. Still, the unsupervised methods scored better in performance than supervised methods and the best discriminative power was obtained when using the LH. For our best performing methods, we identified the top 2 discriminative high-order relationships for LH: (1) (fusiform gyrus, parahyppocampal gyrus) and (Lingual gyrus, pericalcarine cortex), and (2) (entorhinal cortex, transverse temporal gyrus) and (fusiform gyrus, posterior cingulate cortex); along with the top 2 discriminative low-order regions for RH: (1) entorhinal cortex and posterior singulate cortex, and (2) precuneus cortex and postcentral gyrus.

4   Conclusion

In this paper, we presented the first work on a high-order connectomic manifold learning using morphological brain networks for autism identification. Our framework outperformed both supervised and unsupervised baseline methods and was able to further identify the most discriminative relationships between *pairs* of morphological brain connectivities. Noting that ASD classification is a challenging problem, achieving 65.62% is quite promising based on solely T1-w MR images. To improve the connectomic manifold learning for a more accurate ASD/NC segregation, we will evaluate our method on the whole ABIDE dataset, which allow more powerful statistical analysis of our results. Further, we will extend our unsupervised learning method to spatiotemporal connectomic data for monitoring and predicting ASD progression.

References

1. Price, T., Wee, C.-Y., Gao, W., Shen, D.: Multiple-network classification of childhood autism using functional connectivity dynamics. In: Golland, P., Hata, N., Barillot, C., Hornegger, J., Howe, R. (eds.) MICCAI 2014. LNCS, vol. 8675, pp. 177–184. Springer, Cham (2014). doi:10.1007/978-3-319-10443-0_23
2. Koshino, H., Carpenter, P.A., Minshew, N.J., Cherkassky, V.L., Keller, T.A., Just, M.A.: Functional connectivity in an fMRI working memory task in high-functioning autism. Neuroimage **24**, 810–21 (2005)
3. Uddin, L.Q., Supekar, K., Lynch, C.J., Khouzam, A., Phillips, J., Feinstein, C., Ryali, S., Menon, V.: Salience network-based classification and prediction of symptom severity in children with autism. JAMA Psychiatry **70**, 869–879 (2013)
4. Liu, M., Du, J., Jie, B., Zhang, D.: Ordinal patterns for connectivity networks in brain disease diagnosis. In: Ourselin, S., Joskowicz, L., Sabuncu, M.R., Unal, G., Wells, W. (eds.) MICCAI 2016. LNCS, vol. 9900, pp. 1–9. Springer, Cham (2016). doi:10.1007/978-3-319-46720-7_1

5. Ghanbari, Y., Smith, A.R., Schultz, R.T., Verma, R.: Connectivity subnetwork learning for pathology and developmental variations. In: Mori, K., Sakuma, I., Sato, Y., Barillot, C., Navab, N. (eds.) MICCAI 2013. LNCS, vol. 8149, pp. 90–97. Springer, Heidelberg (2013). doi:10.1007/978-3-642-40811-3_12

6. Jbabdi, S., Johansen-Berg, H.: Tractography: where do we go from here? Brain Connect **1**, 169–183 (2012)

7. Smith, E., Thurm, A., Greenstein, D., Farmer, C., Swedo, S., Giedd, J., Raznahan, A.: Cortical thickness change in autism during early childhood. Hum. Brain Mapp. **37**, 2616–2629 (2016)

8. Chen, X., Zhang, H., Shen, D.: Ensemble hierarchical high-order functional connectivity networks for MCI classification. In: Ourselin, S., Joskowicz, L., Sabuncu, M.R., Unal, G., Wells, W. (eds.) MICCAI 2016. LNCS, vol. 9901, pp. 18–25. Springer, Cham (2016). doi:10.1007/978-3-319-46723-8_3

9. Brown, C., Hamarneh, G.: Machine learning on human connectome data from MRI, arXiv:1611.08699v1 (2016)

10. Iidaka, T.: Resting state functional magnetic resonance imaging and neural network classified autism and control. Cortex **63**, 55–67 (2014). Elsevier

11. Li, H., Xue, Z., Ellmore, T.M., Frye, R.E., Wong, S.T.: Identification of faulty DTI-based sub-networks in autism using network regularized SVM. In: IEEE ISBI (2012)

12. Wang, X., Sontag, D., Wang, F.: Unsupervised learning of disease progression models. In: Proceedings of the KDD 2014, pp. 85–94 (2014)

13. Gao, H., et al.: Identifying connectome module patterns via new balanced multi-graph normalized cut. In: Navab, N., Hornegger, J., Wells, W.M., Frangi, A.F. (eds.) MICCAI 2015. LNCS, vol. 9350, pp. 169–176. Springer, Cham (2015). doi:10. 1007/978-3-319-24571-3_21

14. Chen, H., Iraji, A., Jiang, X., Lv, J., Kou, Z., Liu, T.: Longitudinal analysis of brain recovery after mild traumatic brain injury based on groupwise consistent brain network clusters. In: Navab, N., Hornegger, J., Wells, W.M., Frangi, A.F. (eds.) MICCAI 2015. LNCS, vol. 9350, pp. 194–201. Springer, Cham (2015). doi:10. 1007/978-3-319-24571-3_24

15. Wang, B., Zhu, J., Pierson, E., Ramazzotti, D., Batzoglou, S.: Visualization and analysis of single-cell RNA-SEQ data by kernel-based similarity learning. Nature **70**, 869–79 (2017)

16. Maaten, L., Hinton, G.: Visualizing data using t-SNE. J. Mach. Learn. Res. **9**, 2579–2605 (2008)

17. Mueller, S.G., Weiner, M.W., Thal, L.J., Petersen, R.C., Jack, C., Jagust, W., Trojanowski, J.Q., Toga, A.W., Beckett, L.: The Alzheimer's disease neuroimaging initiative. Neuroimaging Clin. North Am. **10**, 869–877 (2005)

18. Joe, H., Ward, J.: Hierarchical grouping to optimize an objective function. J. Am. Stat. Assoc. **58**, 236–244 (1963)

A Unified Bayesian Approach to Extract Network-Based Functional Differences from a Heterogeneous Patient Cohort

Archana Venkataraman[1(✉)], Nicholas Wymbs[2,3], Mary Beth Nebel[2,3], and Stewart Mostofsky[2,3,4]

[1] Department of Electrical and Computer Engineering,
Johns Hopkins University, Baltimore, USA
`archana.venkataraman@jhu.edu`
[2] Center for Neurodevelopmental Medicine and Research, Kennedy Krieger Institute,
Johns Hopkins Scool of Medicine, Baltimore, USA
[3] Department of Neurology, Johns Hopkins School of Medicine, Baltimore, USA
[4] Department of Pediatrics, Johns Hopkins School of Medicine, Baltimore, USA

Abstract. We present a generative Bayesian framework that automatically extracts the hubs of altered functional connectivity between a neurotypical and a patient group, while simultaneously incorporating an observed clinical severity measure for each patient. The key to our framework is the latent or hidden organization in the brain that we cannot directly access. Instead, we observe noisy measurements of the latent structure through functional connectivity data. We derive a variational EM algorithm to infer both the latent network topology and the unknown model parameters. We demonstrate the robustness and clinical relevance of our model on a population study of autism acquired at the Kennedy Krieger Institute in Baltimore, MD. Our model results implicate a more diverse pattern of functional differences than two baseline techniques, which do not incorporate patient heterogeneity.

1 Introduction

Functional connectomics explores the intrinsic organization of the brain via the underlying assumption that two regions, which reliably co-activate are more likely to participate in the same neural processes than two uncorrelated or anti-correlated regions [1]. It has become ubiquitous in the study of neurological disorders, such as schizophrenia and autism. From a practical standpoint, these functional relationships are typically evaluated in resting-state fMRI (rsfMRI), which does not require patients to complete challenging experimental paradigms. Neuroscientifically, group-level changes in the functional architecture of the brain are treated as biomarkers of a particular neurological condition.

State-of-the-art methods follow a two-step procedure of first fitting a connection- or graph-based model and then identifying group differences. Unfortunately, connection-based effects [2] are difficult to interpret and nearly impossible to verify through direct stimulation. While large-scale graph properties,

© Springer International Publishing AG 2017
G. Wu et al. (Eds.): CNI 2017, LNCS 10511, pp. 60–69, 2017.
DOI: 10.1007/978-3-319-67159-8_8

such as modularity [3] and small-worldness [4], mitigate these limitations, are markedly removed from the original network and rarely illuminate a concrete etiological mechanism. Additionally, most studies implicitly treat the patient group as homogeneous, for example, by conducting a statistical evaluation that differentiates patients from controls. This simplification has likely contributed to the lack of reproducible rsfMRI findings in the clinical literature [5].

This paper tackles a fundamental yet overlooked question in the study of functional connectomics: how do we identify the altered functional pathways given a heterogeneous patient cohort? Going one step beyond conventional graph analytics, we will characterize *the full network topology*, i.e., the entire collection of nodes (brain regions) and edges (functional connections) associated with the affected subnetwork. Our framework is based on two guiding principles: (1) complex neurological disorders reflect a distributed but interrelated network of functional impairments, (2) the influence of this affected subnetwork is moderated by the observed clinical severity. Hence, rather than dismissing or regressing out the clinical scores, these measures will crucially guide our network estimation procedures. We draw from the Bayesian model of [6]; however, our novel data likelihood reflects the patient-specific contributions of two functional templates.

We evaluate our model on a population study of Autism Spectrum Disorder (ASD). ASD is characterized by impaired social-communicative skill and awareness across multiple sensory domains, coupled with restricted/repetitive behaviors. Despite ongoing efforts, the complex and heterogeneous presentation of ASD has impeded the discovery of robust neuroimaging biomarkers for the disorder. Functional connectomics has largely implicated the default mode [7] and large-scale network measures [2]. However, these approaches blur information across regions and connections, so it is unclear what neural processes are being impacted. In contrast, our mathematical framework will automatically infer the altered functional pathways, as informed by autism severity.

2   Generative Model of Abnormal Communities

We hypothesize that a given neurological disorder reflects *coordinated disruptions* in the brain. Although we cannot specify *a priori* where these disruptions will occur, we assume that the affected regions will communicate differently with other parts of the brain than if the disorder were not present. In the functional connectomics realm, our assumption can be modeled by region hubs, which exhibit a large number of altered functional connections, as compared to the neurotypical cohort. Below, we refer to these region hubs as *disease foci*; the altered connectivity pattern is termed the *canonical network*.

Following the methodology of [6], we define latent functional connectivity templates F_{ij} and $\bar{F}_{ij}$, which capture the neural synchrony between region i and region j in the neurotypical (i.e., control) and clinical populations, respectively. Empirically, we find that three states: low ($F_{ij} = 0$), medium ($F_{ij} = 1$), and high ($F_{ij} = 2$), best capture the dynamic range and variability of our data. The rsfMRI correlation B_{ij}^l for control subject l is a noisy observation of the latent

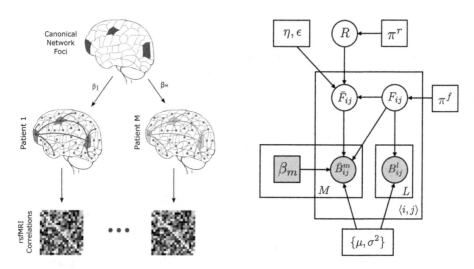

Fig. 1. Hierarchical network model. **Left:** Conceptual diagram of behavioral influence. Red regions correspond to the disease foci, and red edges specify the canonical functional network. Green edges are normal (i.e., healthy) connections. The canonical network contribution for each patient m is specified by the clinical severity, $\beta_m \in [0,1]$. Here, $\beta_1 > \beta_M$, as indicated by the darker edges. **Right:** Graphical model representation. The label R_i indicates whether region i is healthy or abnormal. The neurotypical template $\{F_{ij}\}$ provides a baseline functional architecture for the brain, whereas the clinical template $\{\bar{F}_{ij}\}$ describes the canonical network organization. The patient rsfMRI correlations $\{\bar{B}_{ij}^m\}$ are generated according to the clinical scores $\{\beta_m\}$. (Color figure online)

template F_{ij}. However, the rsfMRI correlations $\{\bar{B}_{ij}^m\}$ for patient m are drawn from either latent template in proportion to the observed clinical severity $\beta_m \in [0,1]$. Figure 1 outlines the full generative process.

Our discrete representation of latent functional connectivity is a notable departure from conventional analysis. Essentially, we assume that the rsfMRI correlations fall into one of three general categories, and that differences in the *bin assignments* are the relevant markers of a disorder. The beauty of our framework is that we isolate the disorder-induced effects in the latent structure, while accommodating noise and subject variability via the data likelihood.

Disease Foci: The binary variable R_i indicates whether region i is healthy ($R_i = 0$), or whether it is a disease foci ($R_i = 1$). We assume an i.i.d. Bernoulli prior: $P(R_i = 1; \pi^r) = \pi^r$. The unknown parameter π^r is shared across regions.

Latent Network Topology: The latent functional connectivity F_{ij} denotes the co-activation between regions i and j in the neurotypical template. Once again, F_{ij} is modeled as a tri-state random variable with an i.i.d. multinomial prior across all pairwise connections: $P(F_{ij} = s; \pi^f) = \pi_s^f, \forall s = 0, 1, 2$.

The clinical template $\bar{F}_{ij}$ depends on both the neurotypical template F_{ij} and the region labels R. We define this variable via three simple rules: (1) a connection between two disease foci is abnormal, (2) a connection between two healthy regions is normal, and (3) a connection between a healthy region and a disease foci is abnormal with unknown probability η. Ideally, $\bar{F}_{ij} = F_{ij}$ for healthy connections, and $\bar{F}_{ij} \neq F_{ij}$ for abnormal connections. However, to better accommodate noise, we allow the clinical template to deviate from these rules with probability ϵ. Mathematically, the conditional distribution is given by

$$
P(\bar{F}_{ij}|F_{ij}, R_i, R_j, \eta, \epsilon) = \begin{cases} \epsilon^{F_{ij}^{\mathrm{T}} \bar{F}_{ij}} \left(\frac{1-\epsilon}{2}\right)^{F_{ij}^{\mathrm{T}} \bar{F}_{ij}}, & R_i = R_j = 1, \\ (1-\epsilon)^{F_{ij}^{\mathrm{T}} \bar{F}_{ij}} \left(\frac{\epsilon}{2}\right)^{F_{ij}^{\mathrm{T}} \bar{F}_{ij}}, & R_i = R_j = 0, \\ \epsilon_1^{F_{ij}^{\mathrm{T}} \bar{F}_{ij}} \left(\frac{1-\epsilon_1}{2}\right)^{F_{ij}^{\mathrm{T}} \bar{F}_{ij}}, & R_i \neq R_j, \end{cases} \tag{1}
$$

where $\epsilon_1 = \eta\epsilon + (1-\eta)(1-\epsilon)$ reflects the interaction between the edge density η and the latent noise ϵ when the region labels differ. For convenience, we have represented the neurotypical connectivity F_{ij} as a length-three binary indicator vector $[F_{ij0} \quad F_{ij1} \quad F_{ij2}]^{\mathrm{T}}$, and likewise for the clinical template.

Data Likelihood: The rsfMRI correlation B_{ij}^l for subject l is generated from a Gaussian distribution, with mean and variance controlled by the neurotypical functional template F_{ij}, i.e., $P(B_{ij}^l|F_{ij} = s; \{\mu, \sigma^2\}) = \mathcal{N}\left(B_{ij}^l; \mu_s, \sigma_s^2\right)$.

In contrast, the patient likelihood weighs the relative contributions of the clinical and neurotypical templates according to the observed severity score $\beta_m \in [0, 1]$. Effectively, the patient rsfMRI correlation $\bar{B}_{ij}^m$ is sampled from a conditional Gaussian mixture with *a priori* probabilities β_m and $1 - \beta_m$.

Using the binary indicator representation for F_{ij} and $\bar{F}_{ij}$, we have

$$
P(\bar{B}_{ij}^m|F_{ij}, \bar{F}_{ij}; \beta_m, \{\mu, \sigma^2\}) = \beta_m \left[\prod_{s=0}^{2} \mathcal{N}\left(\bar{B}_{ij}^m; \mu_s, \sigma_s^2\right)^{\bar{F}_{ijs}}\right]
$$

$$
+ (1 - \beta_m) \left[\prod_{s=0}^{2} \mathcal{N}\left(\bar{B}_{ij}^m; \mu_s, \sigma_s^2\right)^{F_{ijs}}\right]. \tag{2}
$$

Intuitively, patients with larger β_m will more closely follow the clinical template than patients with smaller β_m. The patient-specific analysis in Eq. (2) distinguishes our model from conventional methods and from the prior work of [6].

Variational Inference: We introduce a set of auxiliary random variables $\{Z_{ij}^m\}$, which indicate whether the corresponding rsfMRI measure $\bar{B}_{ij}^m$ is drawn from the clinical ($Z_{ij}^m = 1$) or neurotypical ($Z_{ij}^m = 0$) Gaussian mixture. This strategy allows us to eliminate the sum in Eq. (2) by replacing the conditional density of $\bar{B}_{ij}^m$ with the following joint distribution over Z_{ij}^m and $\bar{B}_{ij}^m$:

$$
P(Z_{ij}^m, \bar{B}_{ij}^m|F_{ij}, \bar{F}_{ij}; \beta_m, \{\mu, \sigma^2\}) = P(Z_{ij}^m; \beta_m)P(\bar{B}_{ij}^m|F_{ij}, \bar{F}_{ij}, Z_{ij}^m; \{\mu, \sigma^2\})
$$

$$
= \left[\beta_m \prod_{s=0}^{2} \mathcal{N}\left(\bar{B}_{ij}^m; \mu_s, \sigma_s^2\right)^{\bar{F}_{ijs}}\right]^{Z_{ij}^m} \left[(1 - \beta_m) \prod_{s=0}^{2} \mathcal{N}\left(\bar{B}_{ij}^m; \mu_s, \sigma_s^2\right)^{F_{ijs}}\right]^{1-Z_{ij}^m} \tag{3}
$$

We combine the above terms to obtain the joint density of latent and observed random variables. Let $\Theta = \{\pi^r, \pi^f, \eta, \epsilon, \mu, \sigma^2\}$ denote the collection of unknown but non-random parameters, and recall that the clinical scores β_m are given. The region labels $\{R_i\}$ induce a complex dependency across pairwise connections $\langle i, j \rangle$. Therefore, we leverage a Variational EM framework to derive the Maximum Likelihood (ML) solution to our model [8].

Our approximate posterior assumes the following factorized form:

$$Q(R, F, \bar{F}, Z) = \prod_{i=1}^{N} q_i^r(R_i; \tilde{\alpha}_i) \prod_{\langle i,j,\rangle} q_{ij}^c(F_{ij}, \bar{F}_{ij}; \tilde{\nu}_{ij}) \prod_{m=1}^{M} \prod_{\langle i,j,\rangle} q_{ij}^z(Z_{ij}^m; \tilde{\gamma}_{ij}^m), \quad (4)$$

where $q_i^r(\cdot)$ and $q_{ij}^z(\cdot)$ are Bernoulli distributions parameterized by $\tilde{\alpha}_i$ and $\tilde{\gamma}_{ij}^m$, respectively. Conversely, $q_{ij}^c(\cdot)$ is a multinomial distribution with 9 states parameterized by $\tilde{\nu}_{ij}$; these states account for the 9 configurations of F_{ij} and $\bar{F}_{ij}$. Equation (4) preserves the connection-wise dependencies in our model while remaining tractable for a large number of regions.

We employ a coordinate descent algorithm to jointly optimize all unknown quantities. During the E-step, we fix Θ and iteratively update the elements of $Q(\cdot)$ to minimize the variational free energy. The updates for $\tilde{\nu}_{ij}$ and $\tilde{\gamma}_{ij}^m$ can be expressed in closed form given the other variational parameters. However, the updates for $\{\tilde{\alpha}_i\}$ are coupled. Therefore, we perform an inner fixed-point iteration until the region posterior converges. In the M-step, we fix $Q(R, F, \bar{F}, Z)$ and optimize the model parameters Θ. The prior and likelihood updates for $\{\pi^r, \pi^f, \mu, \sigma^2\}$ parallel those of a Gaussian mixture model. We then jointly optimize the edge density η and the latent noise ϵ via Newton's method.

Model Evaluation: The marginal posterior $q_i^r(R_i; \tilde{\alpha}_i)$ informs us about the disease foci. We evaluate the robustness of these region assignments via bootstrapping. Specifically, we fit the model to random subsets of the data while preserving the ratio of patients to neurotypical controls. We run two experiments, corresponding to subsets with 90% and 50% of the overall cohort, respectively. Our results are averaged across 100 data re-samplings.

Our canonical network corresponds to the idealized graph of functional differences: $F_{ij} \neq \bar{F}_{ij}$. Despite the confounding latent noise, governed by the parameter ϵ, we can approximate the canonical network based on the max *a posteriori* (MAP) solution for $\{R, F, \bar{F}\}$ and the parameter estimates $\hat{\Theta}$.

Finally, we perform a qualitative comparison of our proposed model with the Bayesian formulation in [6], which assume a homogeneous patient group, and with univariate t-tests on the pairwise rsfMRI correlation coefficients.

Synthetic Experiments: We have run simulations on synthetic data sampled from our model to demonstrate that our variational algorithm can recover the ground truth region labels. Figure 2 illustrates the error in region assignments with respect to two quantities: the latent noise ϵ and the Gaussian separation $\Delta\mu/\sigma$ between adjacent connectivity states assuming equal variances.

In the first experiment (left), we sample disease foci based on the region prior π^r estimated from our autism dataset (see Table 1) and sweep both noise

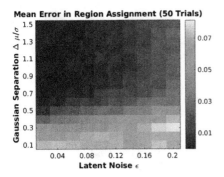
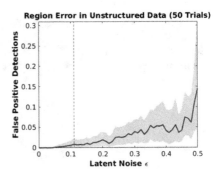

Fig. 2. Probability of error in the inferred region labels, as averaged across 50 generations of synthetic data. The red **X** and red line correspond to the noise regime estimated from our real-world dataset. **Left:** Disease foci were sampled according to π^r in Table 1. **Right:** Uniformly distributed changes in latent functional connectivity. Gray interval denotes the upper and lower standard deviation. (Color figure online)

quantities. In the second experiment (right), we assume that the latent functional differences are uniformly distributed across the brain (i.e., $\pi^r = 0$) and compute the false positive assignments of regions as disease foci. Here, we have fixed the Gaussian separation according to our rsfMRI dataset and focus on the latent noise ϵ. The number of regions, cohort sizes and edge density η are fixed according to the values from our autism dataset. As seen, our algorithm performance is near-perfect for small values of ϵ and larger Gaussian separations. Encouragingly, the region assignment error is small in the noise regime of our real-world dataset, as marked with a red **X** (left) and a red line (right) in Fig. 2.

3   Population Study of Autism

We demonstrate our method on a cohort of 66 children with high-functioning ASD and 66 neurotypical controls, who were matched on the basis of age, gender and IQ. RsfMRI scans were acquired on a Phillips 3T Achieva scanner using a single-shot, partially parallel gradient-recalled EPI sequence ($TR/TE = 2500/30$ ms, flip angle $= 70°$, $res = 3.05 \times 3.15 \times 3$ mm, 128 or 156 time samples). Children were instructed to relax and focus on a cross-hair while remaining still.

RsfMRI preprocessing includes slice time correction, rigid body realignment, and normalization to the EPI version of the MNI template using SPM [9]. The time series were temporally detrended, and we use CompCorr to estimate and remove spatially coherent noise from the white matter and ventricles, along with linearly detrended versions of the six rigid body realignment parameters and their first derivatives [10]. The cleaned data was spatially smoothed (6 mm FWHM Gaussian kernel), temporally filtered using a 0.01–0.1 Hz pass band, and spike-corrected via tools from the AFNI package [11].

We define 116 cortical, subcortical and cerebellar regions based on the Automatic Anatomical Labeling (AAL) atlas [12]. The rsfMRI measure B_{ij}^l is com-

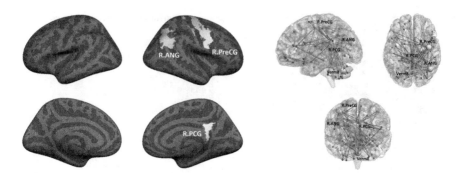

Fig. 3. Results of our heterogeneous patient model. **Left:** Disease foci projected onto the inflated cortical surface. **Right:** Canonical network of abnormal functional connectivity. Yellow nodes correspond to the disease foci. Blue lines signify reduced functional connectivity in ASD; magenta lines denote increased functional connectivity in ASD. (Color figure online)

puted as the Pearson correlation coefficient between the mean time courses of regions i and j. We focus on deviations from baseline by centering the correlation histogram for each subject and fixing $\mu_1 = 0$. Our severity measures β_m correspond to the Autism Diagnostic Observation Schedule (ADOS) total raw score, normalized by the maximum possible test score.

Canonical Network: Figure 3 illustrates the canonical network inferred by our model. The yellow nodes correspond to the disease foci, and we display connections that are consistently implicated across bootstrapping trials. Magenta and blue lines denote increased and reduced latent connectivity in ASD, relative to the neurotypical population. As seen, we identify four disease foci: the right precentral gyrus (R.PreCG), the right posterior cingulate gyrus (R.PCG), the right angular gyrus (R.ANG) and vermis 8 of the cerebellum (Verm8).

Our results are closely aligned with growing evidence, which suggests that brain abnormalities associated with ASD occur at the level of interconnected systems/modules [13,14]. RsfMRI studies in neurotypical subjects have identified several intrinsically connected modules related to visual, motor, auditory, behavioral control, and interoceptive processes [15]. The nodes in Fig. 3 belong to two of these modules: the right precentral gyrus (R.PreCG) and the cerebellar vermis (Verm8) represent critical foci of the sensorimotor network that is specialized in the production of action, while the right posterior cingulate gyrus (R.PCG) and the right angular gyrus (R.ANG) are both key nodes of the default

Table 1. Estimated model parameters for the proposed patient-specific model (top) and the homogeneous model of [6] (bottom).

	π^r	π_0^f	π_1^f	π_2^f	η	ϵ	μ_0	μ_1	μ_2	σ_0^2	σ_1^2	σ_2^2
Prop	0.035	0.28	0.49	0.22	0.16	0.11	−0.18	0.00	0.23	0.037	0.031	0.030
Homogen	0.0087	0.29	0.48	0.22	0.16	0.052	−0.18	0.00	0.22	0.037	0.032	0.031

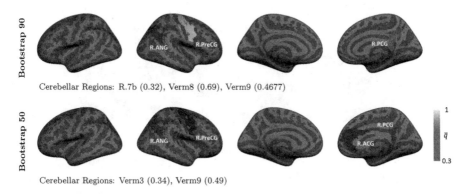

Fig. 4. Average marginal posterior probability $q_i^r(\cdot)$ for each community across 100 random samplings of the rsfMRI dataset. Top row includes 90% of the subjects in each subset, and the bottom row includes 50%. Reproducibility of cerebellar regions are listed underneath. The colorbar denotes the average posterior probability $\bar{q}_i^r$.

mode network (DMN), which is more engaged during self-referential processing and social cognition [16]. Extant ASD research has largely focused on understanding social-communicative deficits in ASD and the potential involvement of the DMN. However, an emerging consensus suggests that movement abnormalities are also specific for ASD [17] and potentially rooted in the intrinsic functional organization of the brain [18]. For example, action execution, imitation, and emulation can be linked to shared functional dynamics between the sensorimotor and DMN systems [19]. As such, communication disruptions between these systems may negatively impact the development of internal action models, which are crucial to both sensorimotor and social skill development in children with ASD [20]. Considered together, our findings support the theory that motor behavior and self-referential processing deficits experienced by individuals with ASD can be jointly attributed to faulty connections within the brain.

Figure 4 reports the average posterior probability $\bar{q}_i^r(\cdot)$ of each region across 100 bootstrapped trials. We display only the regions for which $\bar{q}_i^r > 0.3$ to emphasize the most prominent patterns. As seen, our model consistently recovers the canonical network foci in Fig. 3 when trained on 90% of the data. Remarkably, we are still able to detect the original network foci using half the dataset, which further validates the reproducibility of our Bayesian model. Finally, our bootstrapping experiments also implicate cerebellar regions adjacent to Vermis 8, which ties into broader theories of altered cerebellar functioning in ASD [21].

Figure 5 compares our canonical network (left) with the model of [6] (middle), which assumes a homogeneous patient group, and with standard univariate tests (right). Notice that while the estimated model parameters in Table 1 are nearly identical, the proposed and homogeneous Bayesian models implicate different functional networks. Specifically, the homogeneous model identifies a single disease foci (R.ANG). However, incorporating the severity scores β_m seems to provide an additional level of flexibility, which allows us to find robust effects

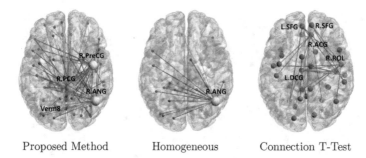

Proposed Method   Homogeneous   Connection T-Test

Fig. 5. Qualitative comparison of our proposed model of patient heterogeneity (left), the original Bayesian model described in [6] (middle), and the top connections ($p <$ 0.001 uncorrected) via two-sample t-tests on the pairwise correlation values (right).

in other brain regions. The connections implicated by two-sample t-tests form a markedly different pattern than the network model results; they tend to concentrate in the frontal cortex and anterior cingulate gyrus. This observation suggests that our disease foci provide a unique perspective of the data.

4   Conclusion

We have introduced a novel probabilistic framework that identifies group differences in functional connectivity while accommodating a heterogeneous clinical presentation. Specifically, we assume a latent graph organization that captures population-level effects. The influence of this latent structure on the data is moderated by the observed clinical severity scores for each patient. Synthetic experiments confirm that our variational algorithm can accurately infer ground-truth region labels under noise levels commiserate to real-world data. We further evaluate our model on a population study of high-functioning ASD. Our results implicated a distributed network of abnormal connectivity that concentrates in the precentral gyrus, posterior cingulate, angular gyrus and cerebellar vermis. We use bootstrapping to verify the robustness of our region assignments, and we demonstrate that our model identifies a richer set of functional differences than two baseline approaches, which do not account for patient heterogeneity.

Acknowledgments. This work was supported in part by the National Institute of Mental Health (R01 MH085328-09, R01 MH078160-07, and K01 MH109766), the National Institute of Neurological Disorders and Stroke (R01 NS048527-08), and the Autism Speaks foundation.

References

1. Fox, M.D., Raichle, M.E.: Spontaneous fluctuations in brain activity observed with functional magnetic resonance imaging. Nature **8**, 700–711 (2007)

2. Di Martino, A., Yan, C.G., Li, Q., Denio, E., Castellanos, F., et al.: The autism brain imaging data exchange: towards a large-scale evaluation of the intrinsic brain architecture in autism. Mol. Psychiatry **19**, 659–667 (2014)
3. Meunier, D., et al.: Hierarchical modularity in human brain functional networks. Front. Neuroinform. **3**, 37 (2009)
4. Sporns, O., Zwi, J.: The small world of the cerebral cortex. Neuroinformatics **2**, 145–162 (2004)
5. Horga, G., Kaur, T., Peterson, B.: Annual research review: current limitations and future directions in MRI studies of child- and adult-onset developmental psychopathologies? J. Child Psychol. Psychiatry **55**, 659–680 (2014)
6. Venkataraman, A., et al.: From brain connectivity models to region labels: Identifying foci of a neurological disorder. IEEE TMI **32**, 2078–2098 (2013)
7. Padmanabhana, A., et al.: The default mode network in autism. Biol. Psychiatry Cogn. Neurosci. Neuroimaging, 1–11 (2017, in press)
8. Jordan, M., Ghahramani, Z., Jaakkola, T.S., Saul, L.K.: An introduction to variational methods for graphical models. Mach. Learn. **37**, 183–233 (1999)
9. Friston, K., et al. (eds.): Statistical Parametric Mapping: The Analysis of Functional Brain Images. Academic Press, London (2007)
10. Behzadi, Y., et al.: A component based noise correction method (compcor) for bold and perfusion based fMRI. NeuroImage **37**, 90–101 (2007)
11. Cox, R.W.: AFNI: software for analysis and visualization of functional magnetic resonance neuroimages. Comput. Biomed. Res. **29**(3), 162–173 (1996)
12. Tzourio-Mazoyer, N., et al.: Automated anatomical labeling of activations in SPM using a macroscopic anatomical parcellation of the MNI MRI single-subject brain. NeuroImage **15**, 273–289 (2002)
13. Schipul, S., Keller, T., Just, M.: Inter-regional brain communication and its disturbance in autism. Front. Syst. Neurosci. **5**, 10 (2011)
14. Vasa, R., Mostofsky, S., Ewen, J.: The disrupted connectivity hypothesis of autism spectrum disorders: time for the next phase in research. Biol. Psychiatry Cogn. Neurosci. Neuroimaging **1**, 245–252 (2016)
15. Damoiseaux, J., et al.: Consistent resting-state networks across healthy subjects. PNAS **103**, 13848–13853 (2006)
16. Mars, R., et al.: On the relationship between the "default mode network" and the "social brain". Front. Hum. Neurosci. **6**, 189 (2012)
17. Ament, K., et al.: Evidence for specificity of motor impairments in catching and balance in children with autism. J. Autism Dev. Disorders **45**, 742–751 (2015)
18. Nebel, M., et al.: Disruption of functional organization within the primary motor cortex in children with autism. HBM **35**, 567–580 (2014)
19. Bardi, L., et al.: TPJ-M1 interaction in the control of shared representations: new insights from tDCS and TMS combined. NeuroImage **146**, 734–740 (2017)
20. Mostofsky, S., Ewen, J.: Altered connectivity and action model formation in autism. Neuroscientist **17**, 437–448 (2011)
21. Becker, E., Stoodley, C.: Autism spectrum disorder and the cerebellum. Int. Rev. Neurobiol. **113**, 1–34 (2013)

FCNet: A Convolutional Neural Network for Calculating Functional Connectivity from Functional MRI

Atif Riaz$^{(\boxtimes)}$, Muhammad Asad, S.M. Masudur Rahman Al-Arif,
Eduardo Alonso, Danai Dima, Philip Corr, and Greg Slabaugh

City, University of London, London, UK
atif.riaz@city.ac.uk

Abstract. Investigation of functional brain connectivity patterns using functional MRI has received significant interest in the neuroimaging domain. Brain functional connectivity alterations have widely been exploited for diagnosis and prediction of various brain disorders. Over the last several years, the research community has made tremendous advancements in constructing brain functional connectivity from time-series functional MRI signals using computational methods. However, even modern machine learning techniques rely on conventional correlation and distance measures as a basic step towards the calculation of the functional connectivity. Such measures might not be able to capture the latent characteristics of raw time-series signals. To overcome this shortcoming, we propose a novel convolutional neural network based model, FCNet, that extracts functional connectivity directly from raw fMRI time-series signals. The FCNet consists of a convolutional neural network that extracts features from time-series signals and a fully connected network that computes the similarity between the extracted features in a Siamese architecture. The functional connectivity computed using FCNet is combined with phenotypic information and used to classify individuals as healthy controls or neurological disorder subjects. Experimental results on the publicly available ADHD-200 dataset demonstrate that this innovative framework can improve classification accuracy, which indicates that the features learnt from FCNet have superior discriminative power.

Keywords: Functional connectivity · CNN · fMRI · Deep learning

1 Introduction

In recent literature, functional magnetic resonance imaging (fMRI) has become a popular neuroimaging modality to explore the functional connectivity (FC) patterns of the brain. Specifically, the resting state FC has shown to reflect a robust functional organization of the brain. Many studies [1–3] have shown promising

© Springer International Publishing AG 2017
G. Wu et al. (Eds.): CNI 2017, LNCS 10511, pp. 70–78, 2017.
DOI: 10.1007/978-3-319-67159-8_9

outcomes in the understanding of brain disorders like schizophrenia, attention deficit hyperactivity disorder (ADHD) and Alzheimer's disease by studying brain functional networks in resting state fMRI. The human brain can be viewed as a large and complicated network in which the regions are represented as nodes and their connectivity as edges of the network. FC is viewed as a pair-wise connectivity measurement which describes the strength of temporal coherence (co-activity) between the brain regions. A number of recent studies have shown FC as an important biomarker for the identification of different brain disorders like ADHD [1], schizophrenia [3] and many more.

Several methods have been developed for extracting the FC from temporal resting state fMRI data such as correlation measures [3], clustering [1] and graph measures [2]. Most of the existing techniques, including modern machine learning methods like clustering, rely on conventional distance-based measures for calculating the strength of similarity between brain region signals. These measures act as hand-crafted features towards determining the FC and, may not be able to capture the inherent characteristics of the time-series signals.

A convolutional neural network (CNN) provides a powerful deep learning model which has been shown to outperform existing hand-crafted features based methods in a number of domains like image classification, image segmentation and object recognition. The strength of a CNN comes from its representation learning capabilities, where the most discriminative features are learned during training. A CNN is composed of multiple modules, where each module learns the representation from one lower level to a higher, more abstract level. To our knowledge, CNNs have not been investigated to determine the FC of brain regions. In this work, our motivation is to construct the FC patterns from fMRI data by exploiting the representation learning capability of a CNN. Particularly, we are interested to determine if a CNN can capture the latent characteristics of the brain signals. Compared with other methods, our approach calculates the FC directly from pairs of raw time-series fMRI signals, naturally preserving the inherent characteristics of the time-series signal in the constructed FC.

For training, FCNet requires pairs of fMRI signals and a real value indicating the degree of FC. Training data is produced using a generator that selects pairs of time-series signals that are considered functionally connected, and those that are not. This data is used to train a Siamese network [4] architecture to predict FC from an input signal pair. We demonstrate the expressive power of the features extracted from the FCNet in a classification framework that classifies individuals as healthy control or disorder subjects.

The proposed framework has several stages and is illustrated in Fig. 1. The first stage is to train the proposed FCNet using the data generated by a data generator (Fig. 1a). The FCNet learns to infer the FC between the brain regions. Once the FCNet is trained, the next step is to use the FCs to distinguish healthy control and disorder subjects. This is accomplished by the classification pathways (Fig. 1b, c). During training, the fMRI signal from a training subject is fed into the trained FCNet, which generates a FC map of the brain regions. Then an Elastic Net (EN) [5] is used to extract the most discriminative features from

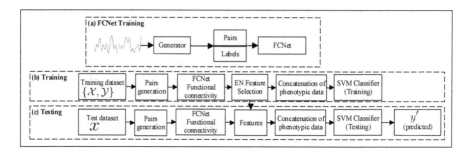

Fig. 1. Flowchart of the proposed method. In (a), FCNet is trained from the data generated by the generator. In the training pipeline (b), functional connectivity (FC) is generated through FCNet. Next, discriminant features are selected and are concatenated with phenotypic data, then employed to train a SVM classifier. The testing pipeline is shown in (c). After FC is calculated, features are selected and concatenated with phenotypic data. A trained SVM is employed for classification.

the FC. The process combines variable shrinkage and grouped feature selection. These features are concatenated with phenotypic information to create a final feature map. The feature map is used to train a SVM classifier which learns to classify between healthy control and disorder subjects. Once the classification path of Fig. 1b is trained, it can be used to classify test subjects as shown in Fig. 1c.

The contributions of this work include: (1) a novel CNN-based deep learning model for extraction of functional connectivity from raw fMRI signals (2) a learnable similarity measure for calculation of functional connectivity and (3) improved classification accuracy over the state-of-the-art on the ADHD-200 dataset.

2   Method

2.1   Data and Preprocessing

The resting state fMRI data evaluated in this work is from the ADHD-200 consortium [6]. Different imaging sites contributed to the dataset. The data is comprised of resting state functional MRI data as well as phenotypic information. The consortium has provided a training dataset, and an independent testing dataset separately for each imaging site. We have used data from three sites: NeuroImage (NI), New York University Medical Center (NYU) and Peking University (Peking). All sites have a different number of subjects. Additionally, imaging sites have different scan parameters and equipment, which increases the complexity and diversity of the dataset. This data has been preprocessed as part of the connectome project[1] and brain is parcellated into 90 regions using the automated anatomical labelling atlas [7]. A more detailed description of the

---

[1] www.preprocessed-connectomes-project.org/adhd200/.

data and pre-processing steps appears on the connectome website. We have integrated phenotypic information of age, gender, verbal IQ, performance IQ and Full4 IQ for NYU and Peking (for NeuroImage, phenotypic information of IQs was not available).

2.2 Functional Connectivity Through FCNet

In this work, we propose a novel deep CNN for the calculation of FC. Our proposed method calculates FC directly from raw time-series signals instead of relying on conventional similarity measures like correlation or distance based measures.

FCNet is a deep-network architecture for jointly learning a feature extractor network that captures the features from the individual regional time-series signal and a learnable similarity network that calculates similarity between the pairs. The FCNet is presented in Fig. 2 and individual networks are detailed below.

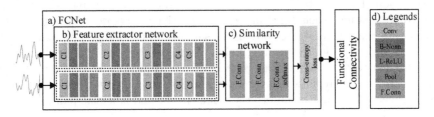

Fig. 2. Architecture of the FCNet. (a) FCNet with coupled feature extractor network (one network for each brain region) and the similarity network which measures the degree of similarity between the two regions. (b) The feature extractor network which includes multiple layers namely Convolutional (Conv), Batch Normalization (B-Norm), Pooling (pool), Fully Connected (F.Conn) and Leaky-ReLU (L-ReLU). (c) The similarity measure network. (d) Legends for feature extractor network.

The Feature Extractor Network: This network extracts features from *individual* brain region time-series signals and is comprised of multiple layers that are common in CNN models to learn abstract representations of features. Here, we use a Leaky Rectified Linear Unit (ReLU) as the non-linearity function, due to its faster convergence over ReLU [8]. The network accepts time-series signal of length 172. All pooling layers pool spatially with pool length of 2. For all convolution layers, we use kernel size of 3 and the number of filters are 32, 64, 96, 64, 64 for layers $C1$, $C2$, $C3$, $C4$, $C5$ respectively. The last fully connected layer in the network has 32 nodes.

The Similarity Measure Network: This network employs a neural network to learn the FC between *pairs* of extracted features from two brain regions. This is in contrast to conventional methods that use hand-crafted computations like correlation or distance based measures. The input to this network are the abstracted features extracted from two regions. The network computes their FC, which relates to the similarity between the two regions. The network is comprised of three fully connected layers where the last layer is connected to a softmax classifier with dense connections. Next, we describe architectural considerations and training.

Coupled Architecture with Shared Parameters: In order to calculate the FC between different pairs of brain regions, the brain regions must undergo the same feature extraction processing. It can be realized by employing the two feature extractor networks (coupled structure) with the constraint that both networks share the same set of parameters. During the training phase, updates are applied to the shared parameters. The approach is similar to Siamese network [4] that is used to measure similarity between two images.

Data Generator for Training FCNet: For training FCNet, we require similar (functionally connected) and dissimilar (not functionally connected) regions with corresponding labels (one and zero respectively). We develop a generator to generate pairs of brain regions using support from affinity propagation [9] clustering for labelling the training pairs. We make pairs for regions that lie in the same cluster and assign them the label one (functionally connected). For unconnected pairs (regions that are not functionally connected), we randomly pick regions that do not belong to the same cluster and label the pair zero. The procedure is detailed in Algorithm 1.

Algorithm 1. Data generation for training of the FCNet.

Input: X % **X** is the subjects in training data, nReg (number of regions) = 90.
Output: (Pairs, Labels) % Pairs and Labels are used for training of FCNet.
1 **for each** x *in* **X do**
2     $c \leftarrow \text{cluster}(x)$ % clustering results in c
3   count $\leftarrow 0$
4   **for** $i \leftarrow 1$ *to* *nReg* **do**
5   **for each** j *in* $(1 \rightarrow nReg)$ *such that* $c(x_i) = c(x_j)$ *and* $i \neq j$ **do**
6   AddToPairs((x_i, x_j), Pairs)
7   AddToLabels(1, Labels)
8   count $\leftarrow$ count $+ 1$
9   **end**
10   **for** $k \leftarrow 1$ *to* *count* **do**
11           $r \leftarrow \text{RandomSelectRegion}(x)$ such that $c(x_i) \neq c(r)$
12   AddToPairs((x_i, r), Pairs)
13   AddToLabels(0, Labels)
14   **end**
15   **end**
16 **end**
17 return (Pairs,Labels)

Training of FCNet: FCNet is trained on pair-wise signals with labels generated from the generator as described above. The FCNet is trained end-to-end using a coupled architecture minimizing the cross-entropy loss

$$L_{fc} = -\frac{1}{n}\sum_{1}^{n}[y_i log(\hat{y_i}) + (1 - y_i)log(1 - \hat{y_i})], \tag{1}$$

where n is the number of training samples, y_i is the label of pairs (1 for functionally connected and 0 for unconnected regions) and $\hat{y_i}$ is the prediction by the softmax layer.

To evaluate FC through the FCNet, regions belonging to each subject are grouped into pairs (for 90 regions belonging to a subject 4005 unique pairs are created). The pairs are passed to the trained FCNet, which computes FC for each pair.

2.3   Feature Selection and Classification

The FC of a subject may contain highly correlated features. We investigate Elastic Net (EN) based feature selection [5] for extracting discriminant features. EN combines the L_1 penalty to enable variable selection and continuous shrinkage, and the L_2 penalty to encourage grouped selection of features. If $\boldsymbol{y}$ is the label vector for subjects $y_i \epsilon (l_1, l_2, ...l_n)$ and $\boldsymbol{X} = \{FC_1, FC_2, ...FC_n\}$ represents the functional connectivity of subjects, we minimize the cost function

$$L_{en}(\lambda_1, \lambda_2, \beta) = (||\boldsymbol{y} - \boldsymbol{X}\beta||)^2 + \lambda_1(||\boldsymbol{\beta}||)_1 + \lambda_2||\boldsymbol{\beta}||^2, \tag{2}$$

where λ_1 and λ_2 are weights of the terms forming the penalty function, and β coefficients are calculated through model fitting. The features with non zero β coefficients relating to minimum cross validation error are extracted. Similar to [1], phenotypic information of the subjects are concatenated with the EN based selected features to construct a combined feature set for classification.

The final step in the proposed framework is classification where a support vector machine (SVM) classifier is utilized to evaluate the discriminative ability of the selected features.

3   Experiments and Results

The proposed framework is evaluated on a dataset provided by the ADHD-200 consortium, and contains four categories of subjects: controls, ADHD combined, ADHD hyperactive-impulsive and ADHD inattentive. Here we combine all ADHD subtypes in one category since we want to investigate classification between healthy control and ADHD.

In many biomedical domains specifically fMRI, scarcity of the data emerges as a challenging task. To address this issue, we combine all subjects from training

Table 1. Comparison of FCNet with the average results of competition teams, highest accuracy achieved for individual site, correlation based FC and state-of-the-art clustering based FC results [1]. The highest accuracy for NI was not quoted by [10].

	NI	Peking	NYU
Average accuracy [6]	56.9%	51.0%	35.1%
Highest accuracy [10]	–	58%	56%
Clustering method [1]	44%	65%	61%
Correlation	52.0 %	52.9%	56.1%
Proposed method	**64.0%**	**68.6%**	**63.4%**

datasets of the different imaging sites and FCNet is trained on this combined training dataset. Feature selection and classification is evaluated on individual imaging datasets. The trained SVM classifier is tested with independent test data provided for each individual site, and results are presented in Table 1. The results show that our method outperforms the average accuracy results of competition teams (data from the competition website), highest accuracy for any individual site (from [10]) and correlation-based FC results. For correlation based results, FC is calculated through correlation and the rest of processing pipeline is same as our method. It is worth noting that the parameters of our framework are held constant for all the imaging datasets. Our method also performs well in comparison with a state-of-the-art clustering based FC technique [1]. In order to compare with the related work [1] that employed phenotypic information, we compare and present the results in Table 2, which shows that our method performs well in all of the three imaging sites. Finally, in order to study the FC differences between the healthy control group and the ADHD group, we visualize their respective FC patterns using the Peking dataset and present the results in Fig. 3. The results show that in ADHD, the temporal lobe functional connectivity is reduced compared to healthy controls.

Table 2. Comparison of proposed method with the state-of-the-art results [1]. The results suggest that the FCNet outperforms the state-of-the-art classification accuracy.

Phenotypic information	Method	NI	Peking	NYU
Not used	Clustering method [1]	44%	58.8%	24.3%
	Proposed method	60.0%	62.7%	58.5%
Used	Clustering method [1]	–	65%	61%
	Proposed method	**64.0%**	**68.6%**	**63.4%**

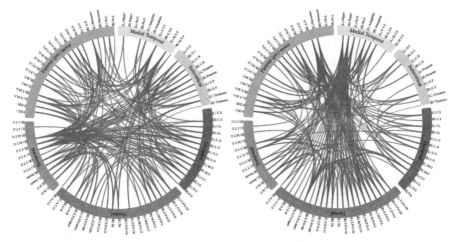

(a) FC patterns of the healthy control group.   (b) FC patterns of the ADHD group.

Fig. 3. Comparison of mean functional connectivity (FC) of healthy control group (a) and ADHD group (b) for the Peking dataset. For the sake of clarity, only the top 200 connections (based upon their connectivity strength) from both groups are presented. The FC patterns show alterations. The temporal lobe FC patterns are altered the most with a decrease of 15% FC patterns in the ADHD group. The inter-temporal lobe FC patterns are reduced from 22.7% (healthy group) to 7.7% (ADHD group).

4   Conclusion

In this paper, we have proposed a novel convolutional neural network-based deep learning model called FCNet for functional connectivity estimation from fMRI data. The proposed model extracts functional connectivity from raw time-series signals instead of relying on any conventional distance based measure. The FCNet is comprised of a feature extractor network that extracts features from the raw time-series signals and a learnable similarity measure network that calculates the similarity between regions. The FCNet is an end-to-end trainable network. After calculating functional connectivity, elastic net is applied to select discriminant features. Finally, a support vector machine classifier is applied to evaluate the classification results. Experimental results on the ADHD-200 dataset demonstrate promising performance with our method.

References

1. Riaz, A., Alonso, E., Slabaugh, G.: Phenotypic integrated framework for classification of ADHD using fMRI. In: Campilho, A., Karray, F. (eds.) ICIAR 2016. LNCS, vol. 9730, pp. 217–225. Springer, Cham (2016). doi:10.1007/978-3-319-41501-7_25
2. Dey, S., Rao, A.R., Shah, M.: Attributed graph distance measure for automatic detection of attention deficit hyperactive disordered subjects. Front. Neural Circuits **8** (2014)

3. Kim, J., Calhoun, V.D., Shim, E., Lee, J.H.: Deep neural network with weight sparsity control and pre-training extracts hierarchical features and enhances classification performance: Evidence from whole-brain resting-state functional connectivity patterns of schizophrenia. NeuroImage **124**, 127–146 (2016)

4. Bromley, J., Guyon, I., LeCun, Y., Säckinger, E., Shah, R.: Signature verification using a "Siamese" time delay neural network. In: Advances in Neural Information Processing Systems, pp. 737–744 (1994)

5. Zou, H., Hastie, T.: Regularization and variable selection via the elastic net. J. R. Stat. Soc. Ser. B (Stat. Methodol.) **67**(2), 301–320 (2005)

6. ADHD-200. http://fcon_1000.projects.nitrc.org/indi/adhd200/

7. Tzourio-Mazoyer, N., Landeau, B., Papathanassiou, D., Crivello, F., Etard, O., Delcroix, N., Mazoyer, B., Joliot, M.: Automated anatomical labeling of activations in SPM using a macroscopic anatomical parcellation of the MNI MRI single-subject brain. NeuroImage **15**(1), 273–289 (2002)

8. Maas, A.L., Hannun, A.Y., Ng, A.Y.: Rectifier nonlinearities improve neural network acoustic models. In: International Conference on Machine Learning, vol. 30 (2013)

9. Frey, B.J., Dueck, D.: Clustering by passing messages between data points. Science **315**(5814), 972–976 (2007)

10. Nuñez-Garcia, M., Simpraga, S., Jurado, M.A., Garolera, M., Pueyo, R., Igual, L.: FADR: functional-anatomical discriminative regions for rest fMRI characterization. In: Zhou, L., Wang, L., Wang, Q., Shi, Y. (eds.) MLMI 2015. LNCS, vol. 9352, pp. 61–68. Springer, Cham (2015). doi:10.1007/978-3-319-24888-2_8

Identifying Subnetwork Fingerprints in Structural Connectomes: A Data-Driven Approach

Brent C. Munsell[1(✉)], Eric Hofesmann[1,2], John Delgaizo[3],
Martin Styner[4], and Leonardo Bonilha[3]

[1] Department of Computer Science, College of Charleston, Charleston, USA
munsellb@cofc.edu
[2] Department of Electrical Engineering and Computer Science,
University of Michigan, Ann Arbor, USA
[3] Department of Neurology, Medical University of South Carolina,
Charleston, USA
[4] Department of Psychiatry, University of North Carolina, Chapel Hill, USA

Abstract. Identifying white matter connectivity patterns in the human brain derived from neuroimaging data is an important area of research in computational medicine. Recently, machine learning techniques typically use *region-to-region* or *hub-base* connectivity features to understand how the brain is organized, and then use this information to predict the clinical outcome. Unfortunately, computational models that are trained with these types of features are very localized to a particular region in the brain, i.e. one particular brain region or two connected brain regions, and may not provide the level of information needed to understand more complex relationships that span multiple connected brain regions. To overcome this limitation a new *subnetwork* feature is introduced that combine *region-to-region* and *hub-based* delay information using the shortest path algorithm. The proposed feature is then used to construct a deep learning model to recognize the identity of 20 different subjects. The results show person identification models trained with our feature are approximately 30% and 50% more accurate than models trained only using hub-based features and region-to-region features, respectively. Lastly, a *connectome fingerprint* is identified using a neural network backtrack approach that selects the subnetwork features that are responsible for classification performance.

1 Introduction

The brain connectome provides unprecedented information about global and regional conformations of neuronal network architecture (or *network architecture* for brevity) across the entire brain with milimetric precision [1, 2]. Identifying connectivity patterns, commonly called *fingerprints*, in connectome data is a challenging problem. Typically, machine learning algorithms attempt to learn connectivity patterns using features derived from *region-to-region* connectivity information shown in Fig. 1(a), or features derived using a *hub-based* network analysis technique shown in Fig. 1(b). In general, region-to-region information is considered to be a *local* measure of connectivity, in that,

© Springer International Publishing AG 2017
G. Wu et al. (Eds.): CNI 2017, LNCS 10511, pp. 79–88, 2017.
DOI: 10.1007/978-3-319-67159-8_10

it is localized to two connected brain regions. Whereas, hub-based information may be a more *global* measure of connectivity because it considers the entire network topology[1]. In particular, a centrality hub-based measure may quantify how one particular brain region influences communication across subnetwork borders, where a large hub-based measure has more influence, i.e. is critical for communication across subnetwork borders, and a small hub-based measure has less influence, i.e. is not critical for communication across subnetwork borders.

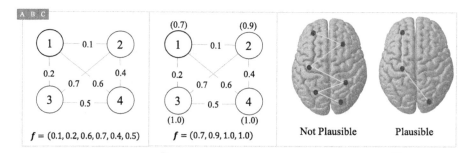

Fig. 1. Example connectivity features derived from (a) region-to-region values (e.g. in graph terminology edge weight values), and (b) hub-based values using the Eigenvector centrality measure annotated in parenthesis above or below the node. (c) Dijkstra's shortest path solution that may not be plausible, and a more biologically plausible shortest path found the proposed modified shortest path algorithm.

Different from previous connectome analysis approaches that only use region-to-region or hub-based features, a new *subnetwork feature* is proposed that combines region-to-region and hub delay information into a single feature that represents a connected subnetwork in the brain. Furthermore, the proposed approach considers the whole brain and is completely *data-driven*, i.e., no prior clinical or anatomical knowledge is used to narrow down which subnetwork features are to be included in the computational model. For instance, in [3] the functional connectome fingerprint that demonstrated the highest classification accuracy was based on two well-known functional subnetworks that are localized to the medial frontal and frontoparietal brain regions, and in [4] the structural connectome fingerprint that demonstrated the highest classification accuracy was based on voxel connectivity values localized to the corpus callosum that is a fiber dense brain region.

Conceptually, our new feature is similar to the spreading dynamic approach introduced in [5], in that, hub regions are important, and are likely shape communication pathways. However, instead of using linear threshold models to understand how these communication pathways spread throughout the brain, our approach uses a modified version of Dijkstra's shortest path algorithm to identify communication pathway backbones. In particular, unlike the traditional shortest path algorithm that only considers region-to-region delay information to compute the shortest path, i.e. the

---

[1] Hub-based may also be a local measure, e.g. node degree or strength are two such examples.

subnetwork that has the least communication delay between a start and end brain region, our modified shortest path algorithm includes a hub-delay based on a centrality hub-based measure.

To overcome issues that may arise between biological plausible mechanisms and topology-based features found using the traditional shortest path algorithm [5], the included hub-delay will likely route communication through brain regions that are critical for communication across subnetworks. In doing so, the resulting subnetwork feature will likely represent a shorter, or simpler, communication path as illustrated in Fig. 1(c). In general, favoring simpler pathways tend to *shrink the topology* of the network, which in turn may represent a more biologically plausible solution.

2 Materials and Methods

2.1 Participants, MRI Acquisition, and Connectome Reconstruction

Twenty subjects were recruited (mean age 34.6 ± 10.66 years) with no history of neurological or psychiatric illnesses. All subject had 3 DWI scans. First scan: 3T Siemens TIM Trio equipped with an 8-channel head coil for signal reception (3D MP-RAGE, TR = 2250 ms, TE = 3.2 ms, 256×256 matrix, 256×256 mm FOV, parallel imaging GRAPPA = 2, 30-directions with b = 1000 s/mm2, TR = 10000 ms, TE = 93 ms, 128×128 matrix parallel imaging GRAPPA = 2, FOV = 243×243 mm, isotropic 1.9 mm voxel size). Second scan: same scanner as the first yielding similar images. The average time between first and second scan was 126.4 ± 102.8 days (range 12–442). Third scan: A different physical unit but same type of MRI scanner (3T Siemens TIM Trio), equipped with a different head coil (12-channel) employing the same imaging sequences. The average time between first scan and third scan was 158.4 ± 103.6 days (range 21–465).

Probabilistic tractography is used estimate the number of white matter streamlines connecting each pair of cortical regions, where the seed regions were obtained through an automatic segmentation process on the T1 weighted images that divided the human cerebral cortex into cortical and subcortical regions of interest (ROIs) based on the Lausanne atlas. This process yielded $m = 83$ ROIs: 41 regions in each hemisphere plus the brainstem. The ROIs were transformed into each subject's DTI space using an affine transformation obtained with FSL's FLIRT. Next, a comprehensive neuronal connectivity matrix, or *connectome*, is calculated for each subject, where connectivity is measured by the number of probabilistic white matter fiber tract streamlines arriving at ROI j when ROI i was seeded, then averaged with the number of probabilistic white matter fiber tract streamlines arriving at ROI i when ROI j was seeded. The resulting *mxm* connectivity matrix C is a symmetric with respect to the main diagonal, and the values are normalized.

2.2 Subnetwork Feature

For each subject an m dimension hub feature vector $\boldsymbol{h}^{\varphi} = (h_1^{\varphi}, \ldots, h_i^{\varphi}, \ldots, h_m^{\varphi})$ is created using the connectivity values in C and the Eigenvector centrality or clustering coefficient hub-based graph-theoretic measures[2] denoted by φ. Next, an $N = m(m-1)/2$ dimension feature vector $\boldsymbol{f}^\varphi = (f_1^\varphi, \ldots, f_\alpha^\varphi, \ldots, f_N^\varphi)$ is created that *combines* the region-to-region connectivity values in the upper diagonal of C with the hub values in $\boldsymbol{h}^\varphi$. More specifically, one feature f_α^φ is a *subnetwork*, or group of connected brain regions, where hub and region-to-region connectivity values both represent communication delays, and $\alpha = (s, t)$ is an index to subscript mapping seen in Fig. 2(c), that defines the start and end brain regions, and $s \neq t$. For example, when $\alpha = N$ the fingerprint start and end brain regions are $(m-1, m)$.

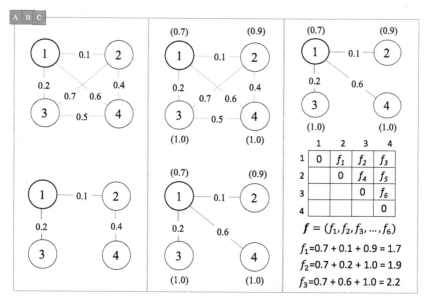

Fig. 2. Examples that illustrates: (a) shortest path algorithm, and (b) modified version of the shortest path algorithm that combines path and hub delays. In (a) and (b) the start node is 1, and in (b) the normalized Eigenvector centrality hub measure is provided in the parenthesis. Lastly, (c) provides an index to subscript mapping example that is used to calculate the subnetwork features. In addition, example feature calculations are provided when the start node is 1, where a smaller value equals the least delay.

For any start brain region, the shortest path to the remaining $(m-1)$ brain regions are found using a modified version of Dijkstra's shortest path algorithm [6] that is visually illustrated in Fig. 2(a–b). For instance, when node-1 is selected to be the start node, the family of shortest paths found by original version of Dijkstra's algorithm are shown in Fig. 2(a). On the other hand, the family of shortest paths shown in Fig. 2(b)

---

[2] The hub-based measures are computed using the publically available brain connectivity toolbox (https://sites.google.com/site/bctnet/).

are those found by the modified version, which includes the Eigenvector centrality hub delay. As seen in Fig. 2(b), when the hub delay is included in the shortest path calculation, the solution will favor simpler pathways, i.e. number of nodes between the start node and stop node is less.

In particular, similar to the original version, the modified version of shortest path algorithm incrementally updates two values: (1) path delay d_u at brain region u, and (2) the predecessor p_u brain region at brain region u. However, a hub delay cost h_v is added

$$\text{if } d_v + \frac{1}{log(C(v,u))} + \frac{1}{log(h_v)} < d_u \textbf{ then } d_u \leftarrow d_v + \frac{1}{log(C(v,u))} + \frac{1}{log(h_v)}; \quad (1)$$
$$p_u \leftarrow v,$$

where $C(u,v)$ is the connectivity value from brain region u to brain region v and $log(\cdot)$ is the natural logarithm. More precisely, the new feature

$$f_\alpha^\varphi = d_t + \frac{1}{log(h_t)}, \text{and } \alpha = (s,t) \quad (2)$$

defines a subnetwork that is acyclic that measures the total path (d_t) and hub delay (h_t). It is important to note, the natural logarithm is used to ensure, as best as possible, the region-to-region connectivity and hub values are normally distributed[3]. After each subnetwork feature is calculated the inverse value is taken that converts smaller subnetwork features to largest ones, and vice versa.

2.3 Person Identification Model and Performance Evaluation

A nxN training data matrix $F^\varphi = \{f_1^\varphi, f_2^\varphi, \ldots, f_i^\varphi, \ldots, f_n^\varphi\}$ and a $nx20$ dimension subject training label matrix $Y = \{y_1, y_2, \ldots, y_i, \ldots, y_n\}$ are constructed, where row vector $y_i = (y_{1i}, y_{2i}, \ldots, y_{20i})$ defines the binary labels for subnetwork feature vector f_i^φ. Next the subnetwork features in F^φ and the subject label matrix Y are used to train a *person identification model* represented by a dense neural network as shown in Fig. 3, where the hidden-layer architecture is $[3000, 1000, 500, 100, 20]$. One additional supervised learning layer is added when the model is trained that also has 20 nodes, one for each training subject. Based primarily on the small size of the training population, the back-propagation optimization procedure uses the stochastic gradient decent algorithm [7] and the categorical cross-entropy loss function [8] to compute the edge weight and bias values at each layer that result in the highest classification accuracy. Once the supervised training step completes, the supervised training layer is removed, and the number of nodes in output layer are then used for person classification.

---

[3] Hub value less than one are set to an arbitrarily large number that represents positive infinity.

The classification accuracy of the person identification model is evaluated using a 3-fold cross-validation strategy. Three folds were selected because the data set has 20 different subjects, and each subject had 3 different DTI scans. For each subnetwork feature vector, in each test fold, classification accuracy is evaluated using the known subject labels, where a score of 100% means the identity of all 20 subjects were correctly recognized. The 3-fold cross-validation strategy is repeated 20 times, and the classification accuracy is reported by finding the mean and standard deviation using the results of the 60 randomly generated test folds. Lastly, the optimal momentum and learning rate were found by incorporating a grid search procedure in the cross-validation strategy.

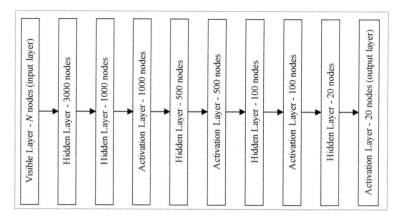

Fig. 3. Dense neural network architecture that defines one visible layer (input layer with N nodes), five hidden layers, and four activation layers, where the last activation layer defines 20 nodes, one for each subject in the connectome data set.

The software used to develop, train, and test the dense neural network was written in Python, and used the publically available Theano[4] and Keras[5] Python deep learning libraries that wrap the C++ NVIDIA CUDA deep neural network libraries[6]. All the reported results were executed on a NVIDIA GeForce GTX 970 graphics card that had 4 GB of memory.

2.4   Feature Selection and Majority Vote Subnetwork Feature

In deep neural network learning approaches selecting input features that have the greatest contribution, and the least contribution, to classification accuracy is a very challenging problem. To address this limitation, we use the backtrack approach similar to that introduced in [9] which starts at the output layer and works backwards through the trained neural network, i.e., through the hidden layers to the input layer (not including the

---

[4] http://deeplearning.net/software/theano/.

[5] https://keras.io/.

[6] https://developer.nvidia.com/cudnn.

activation layers), and follows the nodes that have the largest contribution to the layer directly above. As a result, each input subnetwork feature is assigned a normalized weight value in [0 1], where a value of one implies input subnetwork feature has the greatest contribution to classification accuracy.

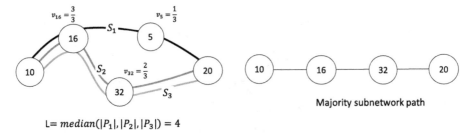

$$L = median(|P_1|, |P_2|, |P_3|) = 4$$

Fig. 4. Given subnetwork feature f_α (start region 10 and end region 20) find the common subnetwork path that represents the brain regions with the greatest occurrence across all three subjects.

Next, for each subnetwork feature selected by the backtrack approach, a majority vote technique is then applied to find the common subnetwork path across all subjects. This step is required because, even though the start and end brain region are the same for one particular subnetwork feature, both the number of nodes in the subnetwork path, i.e. length, and the specific brain regions along the path are likely to differ a small amount between multiple subjects. For instance, given the subnetwork feature f_α that starts at brain region 10 and ends at brain region 20 for subjects S_1, S_2 and S_3 shown in Fig. 4, suppose we wanted to compute the majority subnetwork feature. First, the path length L is identified by finding the median subnetwork path length $|P|$ across each subject. Since we know the start end brain regions of f_α will be the same for each subject, the remaining brain regions that have the greatest occurrence are identified. The top brain region occurrence values are selected, which are then combined with the start and end brain regions to form the *majority subnetwork feature*. Finally, to ensure the proposed approach does not create a majority subnetwork that *does not* exist in any subject, a subnetwork path constraint is included that prevents this degenerate case.

3   Results

At completion of the grid-search, the momentum and learning rate that yielded the reported accuracy results was 0.5 and 0.001, respectively. These optimum model parameter values were then used to determine the classification accuracy summarized in Table 1, and also select the majority subnetwork features that have the greatest influence on classification accuracy. For comparison purposes, the classification performance of five different neural networks were trained and tested using: (1) the proposed majority subnetwork features, (2) hub-only features, and (3) region-to-region connectivity features. Also, the same person identification model neural network

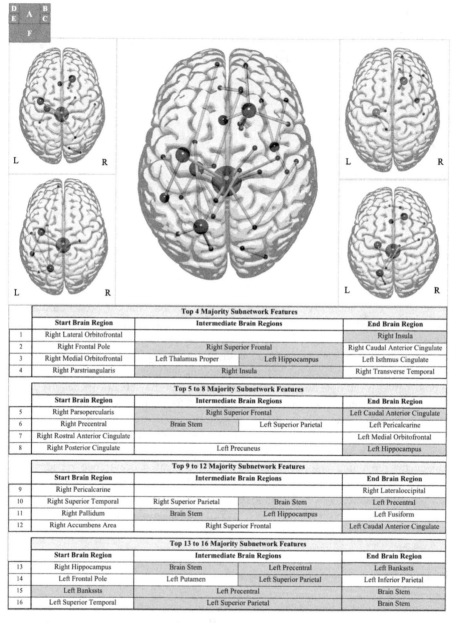

Fig. 5. Subnetwork connectivity diagrams (a) all 16 majority subnetwork features in table (f) that form the connectome fingerprint. Individual subnetworks formed by the (b) top 1-to-4 majority subnetwork features, (c) top 5-to-8 majority subnetwork features, (d) top 9-to-12 majority subnetwork features, and (e) top 13-to-16 majority subnetwork features, also provided in table (f).

Table 1. Mean classification accuracy when the neural network is trained and tested using the five different features. The neural network is trained to recognize the identify of 20 different subjects.

	Classification Accuracy	
	mean (μ)	std deviation (σ)
Proposed subnetwork feature		
Modifed Shortest Path w/ Eigenvector Centrality Hub Measure	0.93	0.04
Modified Shortest Path w/ Clustering Coefficient Hub Measure	0.89	0.06
Hub-based feature		
Clustering Coefficient	0.65	0.09
Eigenvector Centrality	0.57	0.07
Connecotme feature		
Region-to-Region connectivity values	0.41	0.12

design, and optimal model parameter values (momentum and learning rate) were also used in all performance evaluations.

The top 16 majority fingerprint features found by the backtrack feature selection technique and majority vote approach are shown in Fig. 5(f), where table cells highlighted in light blue indicate two, or more, majority subnetwork features that are joined at the same brain region.

Furthermore, the top 16 are categorized into four different groups, namely the top: 1-to-4, 5-to-8, 9-to-12, and 13-to-16. The *connectome fingerprint* defined by the top 16 majority subnetwork features, is shown in Fig. 5(a), where brain regions are represented by red nodes and the size of the node is the related to the number of times the brain region is present in one, or more, majority subnetwork features, e.g., the Brain Stem occurs 6 times. Additionally, the subnetworks formed by the top 1-to-4, 5-to-8, 9-to-12, and 13-to-16 majority subnetworks are shown in Fig. 5(b–e), respectively.

4 Discussion

As shown in Table 1, subnetwork connectivity features that combine pathway and hub delay information, on average, produce trained person identification models that are roughly 91% accurate, whereas models trained only using hub-based features are roughly 62% accurate and models trained using only region-to-region connections are roughly 41% accurate. These results suggest that features that represent subnetworks, based on both hub and pathway delay information, provide a richer descriptor of subtle subnetwork architecture differences that are likely intrinsic to one particular subject. Furthermore, from a neuro-biological point of view, the connectome fingerprint represented by the top 16 majority subnetwork features found by the backtrack approach includes brain regions that are critical to self-awareness, emotion, reward, attention, memory tasks, planning, movement, consciousness, and decision making. All of which would be important to distinguish the identity of an individual.

5 Conclusion

We present a new subnetwork feature that combines hub delay and path delay information using a modified version of Dijkstra's algorithm. Conceptually, the proposed subnetwork feature may allow machine learning techniques to more accurately identify subnetwork patterns and subtle network architecture differences. To assess the performance of our new subnetwork feature, a deep learning classification model is trained using the proposed feature, and was able to recognize the identity of 20 different subjects with 91% accuracy. Finally, a connectome fingerprint is identified using the top 16 majority subnetwork features found by a neural network backtrack approach and the majority vote technique.

Acknowledgement. Would like to thank Dr. Katherine Tom in the Department of Mathematics at the College of Charleston for all the fruitful graph algorithm discussions.

References

1. Sporns, O., Tononi, G., Kotter, R.: The human connectome: a structural description of the human brain. PLoS Comput. Biol. **1**(4), e42 (2005)
2. Sporns, O.: The human connectome: origins and challenges. Neuroimage **80**, 53–61 (2013)
3. Finn, E.S., et al.: Functional connectome fingerprinting: identifying individuals using patterns of brain connectivity. Nat. Neurosci. **18**(11), 1664–1671 (2015)
4. Yeh, F.-C., et al.: Quantifying differences and similarities in whole-brain white matter architecture using local connectome fingerprints. PLoS Comput. Biol. **12**(11), e1005203 (2016)
5. Mišić, B., et al.: Cooperative and competitive spreading dynamics on the human connectome. Neuron **86**(6), 1518–1529 (2015)
6. Cormen, T., et al.: Introduction to Algorithms, 3rd edn. The MIT Press, Cambridge (2009)
7. Bottou, L.: Stochastic gradient learning in neural networks. Proc. Neuro-Nımes **91**(8) (1991)
8. Joe, H.: Relative entropy measures of multivariate dependence. J. Am. Stat. Assoc. **84**(405), 157–164 (1989)
9. Hazlett, H.C., et al.: Early brain development in infants at high risk for autism spectrum disorder. Nature **542**(7641), 348–351 (2017)

A Simple and Efficient Cylinder Imposter Approach to Visualize DTI Fiber Tracts

Lucas L. Nesi[1,3], Chris Rorden[2], and Brent C. Munsell[1(✉)]

[1] Department of Computer Science, College of Charleston, Charleston, USA
munsellb@cofc.edu
[2] Department of Psychology, University of South Carolina, Columbia, USA
[3] Science Without Borders Scholarship/CAPES, Brasilia, Federal District, Brazil

Abstract. The human brain can be divided into two tissue categories, namely: gray matter that maybe associated with cognitive, motor, emotion, and visual processing, and white matter that facilitates neuronal communication between gray matter regions. To better understand the organization of white matter connections in the brain, white matter fiber tracts derived from a diffusion tensor image scan is estimated and visualized by publically available software toolsets. In general, one white matter fiber tract is visualized as a thin 3D cylinder, however this approach has many computational limitations, especially when trying to visualize thousands of fiber tracts that have varying size and length. To overcome this limitation, a very simple and efficient imposter approach is proposed, commonly used in the computer graphics community, that exploits the programmable pipeline architecture found in GPU-based parallel processing systems. Using 10,000 fiber tracts derived from a real DTI scan, we show the rendering speed of our imposter approach is 50% times faster, and requires 900% less memory, when compared visualization approach that uses 3D cylinders.

1 Introduction

White matter fiber tractography from diffusion tensor imaging (DTI) allows clinicians and neuroscientists to understand the white matter connectivity across the entire brain. As shown in Fig. 1(a–b) white matter fiber tractography, or *fiber tracts* for short, are typically visualized using the line primitive commonly found in computer graphics libraries. Because the fiber tract may be hard to see, the line width (e.g. pixel width) of individual fiber tracts are adjusted by clinicians to make the fiber tract easier to recognize, or distinguish from neighboring fiber tracts. At the moment, publically available software tools such as TrackVis[1], DSIStudio[2], and 3DSlicer[3] visualize DTI fiber tracts using line strips or triangle-based cylinders, all of which unfortunately have severe disadvantages. For instance, line strips show poor lighting, and the OpenGL core implementation typically limit smooth lines to a one-pixel thickness. The primary problem with triangulated cylinders is that a complicated surface mesh, like illustrated

---

[1] http://trackvis.org/.
[2] http://dsi-studio.labsolver.org/.
[3] https://www.slicer.org/.

© Springer International Publishing AG 2017
G. Wu et al. (Eds.): CNI 2017, LNCS 10511, pp. 89–97, 2017.
DOI: 10.1007/978-3-319-67159-8_11

in Fig. 1(c), overwhelms the graphics processing unit (GPU), and even small adjustments put a huge burden on the central processing unit (CPU), with the end result being poor responsiveness on desktop systems with dedicated high performance graphics capabilities. Furthermore, GPU capable mobile computing devices (e.g. laptops, tablets, and smartphones) have a limited amount of memory, and compute cores, compared to desktop systems, so the visualization of thousands of fiber tracts using a triangulated cylinder approach cannot be supported with limited hardware capabilities.

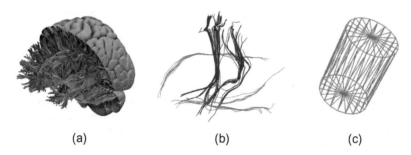

(a)   (b)   (c)

Fig. 1. (a) white matter fiber tracts derived from DTI data visualized using 3D line strips and SurfIce, (b) fiber tracts visualized using line strips that are one-pixel thick, and (c) complexity of triangulated surface mesh that represents a 3D cylinder.

To overcome these software and hardware limitations, our approach applies a common lighting, or *imposter*, trick to a rectangular 3D surface that gives the appearance of a 3D triangulated cylinder. In general, the CPU/GPU benefits of our approach are: (1) Decrease computational complexity, i.e. do not need to create and draw thousands of triangles that define smooth cylinders, and (2) Decrease memory complexity, i.e. does not need to buffer thousands of triangles that define a smooth cylinder surface mesh.

Even though advanced graphic rendering techniques have been applied to visualizing fiber tract data [1, 2] using tubelet, or tuboid, imposter-based frameworks, they are extremely complex, and ultimately extremely difficult to implement. Therefor they are rarely, if at all, used in publically available software toolsets. Unlike the previous tubelet or tuboid imposter-based approaches, our approach is very simple and efficient. More specifically, all vertex and texture information resides in memory on the graphics card, i.e. buffered in memory, which significantly improves memory performance. Additionally, only three vertex points are required to compute the proposed rectangular billboards which are then textured using a buffered normal map to appear like a smooth cylinder surface, this technique significantly improves the computational complexity. As a result, the proposed imposter approach leads to reduced memory consumption and rendering times, and can scale well to systems with limited hardware resources.

2 Proposed Method

The proposed imposter approach is GPU centric, which mean it leverages the parallel programmable graphics pipeline architecture shown in Fig. 2. Using a simple line primitive as an example, i.e. two points, or vertices, connected by a straight line, a shader-based graphics pipeline has three sequential pipeline steps:

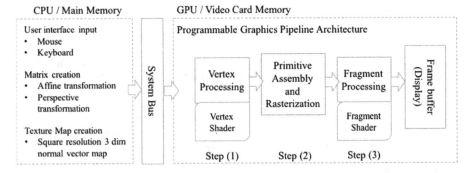

Fig. 2. Schematic diagram that illustrates and basic functions performed by the CPU and GPU, and the programmable graphics pipeline that loads and executes small vertex and fragment shader-based programs related to vertex and fragment processing.

(1) For each vertex, a shader-program is executed that determines how the two vertices should be transformed (e.g. affine transformation matrix created on CPU based on user input from mouse or keyboard),

(2) The transformed vertices are then assembled into a line primitive and rasterized into multiple fragments (e.g. groups of pixels that represent the entire geometric primitive), and

(3) For each fragment, a shader-based program is executed that applies texture (e.g. texture map created on CPU), color, and/or lighting to the pixel values.

Throughout the remainder of this paper, we'll refer to this diagram to indicated were the operation is performed.

For each fiber tract, there are two basic processing steps:

(1) Create a cylinder imposter by applying a normal texture map to rectangular billboard (see Sect. 2.1), and

(2) For only the first and last point on the fiber tract create an end imposter by applying a normal map to a square billboard. This ensure the flat plane appears to be a smooth cylinder for all camera angles (see Sect. 2.2).

2.1 Cylinder Imposter

A fiber tract is initially modeled as a 3D line $(p_1, p_2, \ldots, p_n)$ that is represented by an ordered set of n points like shown in Fig. 3(a), where $p_i = (x_i, y_i, z_i)$ is a point location in a 3D Cartesian coordinate system. Next, for every line segment defined by a fiber tract a 3D rectangular billboard $B = (b_1, b_2, b_3, b_4)$ is created as illustrated in Fig. 3(b), where $b = (x, y, z)$ is a billboard corner point location that are computed on the GPU by the vertex-shader.

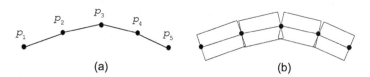

Fig. 3. Basic billboard concept: (a) example white matter fiber tract modeled as a 3D line that is defined by five 3D points and four line segments, and (b) a rectangular 3D billboard is created for each line segment.

More specifically, as seen in Fig. 4(a–b), the vertex-shader first computes three vectors that represent the magnitude and direction of the:

(1) line segment $l = p_{i+1} - p_i$,
(2) camera relative to the start of the line segment $e_i = p_{i+1} - p_{eye}$, and
(3) camera relative to the end of the line segment $e_{i+1} = p_{i+1} - p_{eye}$.

Next, a unit vector $x_i = \frac{l}{\|l\|} \times \frac{e_i}{\|e_i\|}$ that is perpendicular to the start of the line segment, and a unit vector $x_{i+1} = \frac{l}{\|l\|} \times \frac{e_{i+1}}{\|e_{i+1}\|}$ that is perpendicular to the end of the line segment are found, where $\times$ is the cross-product and $\|\cdot\|$ is the L_2 norm. Lastly, using the start and end point locations (p_i, p_{i+1}) and the offset locations vectors (x_i, x_{i+1}), the four corner points are found that define the geometry of the rectangular billboard illustrated in Fig. 4(c).

One drawback of this billboard approach is that a gap between adjacent rectangular billboards will be created as shown in Fig. 4(d). In theory, the largest gap will occur when angle φ between adjacent billboard is 360°. However, such a condition is biologically implausible for white matter fiber tracts, and φ will most likely be between 0 and 90°[4]. To overcome this limitation, in addition to the two points (p_i, p_{i+1}) that define the line segment, the next line segment point p_{i+2} is also provided to the vertex-shader. Using corner points for both line segments, the geometry of the billboard is updated as seen Fig. 4(e) where $b_3 = \hat{b}_1$ and $b_4 = \hat{b}_2$.

Two different matrices that account for affine and perspective operations are computed on the CPU (by user input from mouse or keyboard), and then applied to the rectangular billboard corner points by the vertex-shader on the GPU. Specifically, a 4×4 affine transformation matrix M that is based on the position of the camera using a look-at positioning technique [3], and a 4×4 perspective normalization [3] matrix P that preserves the camera field of view and depth of the geometric object. These two matrices are computed on the CPU and then provided to the vertex-shader.

---

[4] Even 90° is biologically not likely, however this is the upper limit for our implementation.

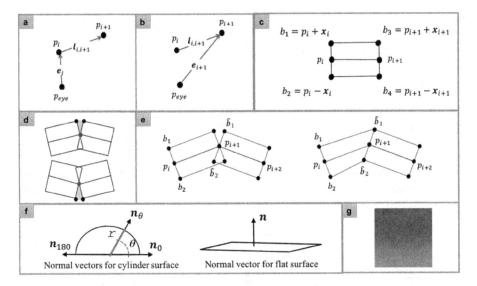

Fig. 4. Example billboard creation for line segment defined by p_i and p_{i+1}: (a) computing the corner point closest to the position of the camera (i.e. the eye point or p_{eye}), (b) computing the corner point farthest from the position of the camera, and (c) the four corner points (b_1, b_2, b_3, b_4). The gap between two adjacent billboards shown in (d) and the fix shown in (e) where corner points $\hat{b}_1 \hat{b}_2$ the derived from the line segment from p_{i+1} to p_{i+2} now become the $b_3 = \hat{b}_1$ and $b_4 = \hat{b}_2$. Normal vector(s) for a cylinder and flat surface shown in (f), and (g) the resulting normal texture map where the coordinate values are converted to RGB.

The orientation of rectangular billboard is updated $\tilde{B} = PMB$ using matrix multiplication to be relative to the camera position.

The fragment-shader then applies a normal texture map to the transformed billboard fragments. At the completion of this step, the flat surface will appear to be a smooth 3D cylinder, however no such cylinder exists. In actuality, this is a lighting trick where a new set of unit normal vectors $n_\theta = (x_\theta, y_\theta, z_\theta)$, i.e. *imposters*, are intentionally created that reflect light like the cylinder surface instead of a flat surface shown in Fig. 4(f). To accomplish this, a $180 \times 180 \times 3$ dimension normal texture map $N(\theta, \phi, \alpha)$ is created, where angle θ is the row index, ϕ is the column index, and α is the axis. For instance, $N(0, 0, 0) = r\cos(0)$ is the x-axis, $N(0, 0, 1) = 0$ is the y-axis, and $N(0, 0, 2) = r\sin(0)$ is the z-axis normal vector values[5] when $\theta = 0$ and $\phi = 0$. An example normal texture map is shown in Fig. 4(g) where the coordinate values have been converted to an RGB image.

Using a linear interpolation technique, the fragment-shader resamples the normal texture map to fit the rectangular billboard fragments, then the vector values fitted to the rectangular billboard are transformed relative to the camera using the inverse

---

[5] In our implementation, the radius r is always equal to 1, and the normal vector is computed relative to the x-z plane.

transpose of the upper-left 3×3 sub-matrix of affine matrix M. Lastly, the Phong lighting technique [4] is applied to the transformed vector normal and the shininess, diffuse, and ambient light components are computed.

2.2   End Imposter

Because the rectangular billboard is a flat plane, if the position of the camera is directly placed in front of, or behind, the rectangular billboard, the fiber tract would appear to be very thin line (i.e. few pixels) as seen in Fig. 5(a). To overcome this limitation, the square billboard in Fig. 5(b) is created and then appended at the beginning of the first line segment, and at the end of the last line segment that defines the entire fiber tract. The construction of the square billboard is quite simple, using the line l and offset

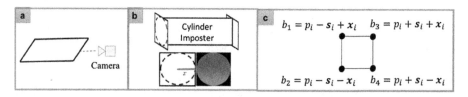

Fig. 5. Example end imposter: (a) camera placement makes billboard appear to be a very thin line, (b) the orientation of the square billboard relative to the rectangular one, and inside the dotted line (described by radius r) the normal map opacity (alpha) values set to one, and outside the dotted line the opacity is set to zero (i.e. invisible), (c) corner points of square billboard.

vectors x_i a new unit offset vector $s_i = \frac{l}{\|l\|} \times \frac{x_i}{\|x_i\|}$ that is perpendicular to billboard plane is found, and the square billboard corner points are computed as illustrated in Fig. 5(c). It should be noted, the same procedure is used for the end of the last line segment, however s_i is calculated using x_{i+1} instead of x_i.

To reduce GPU memory requirements, the buffered normal texture map N applied to rectangular billboard is also applied to the square billboard. However, in order to resemble one-half of a smooth sphere the opacity value of normal vectors outside the radius illustrated in Fig. 5(b) are set to zero making them invisible. Lastly, using the same linear interpolation technique, the buffered normal texture map is fit to the square billboard creating the *end imposter*, then the fitted normal vectors are transformed relative to the camera, and the Phong lighting technique is applied.

3   Experiments

As illustrated in Fig. 6, the visual correctness (that include crossing, curvature and end conditions) of the proposed imposter approach is tested using the OpenGL and WebGL standards on two different data sets: a synthetic 3D helix data set that included four random helix patterns shown in blue, red, yellow, and green, and a real white matter fiber tract data set derived from a DTI scan where the color is related to the length of the fiber tract. In particular, the real data set defines 10,000 different fiber tracts, and the

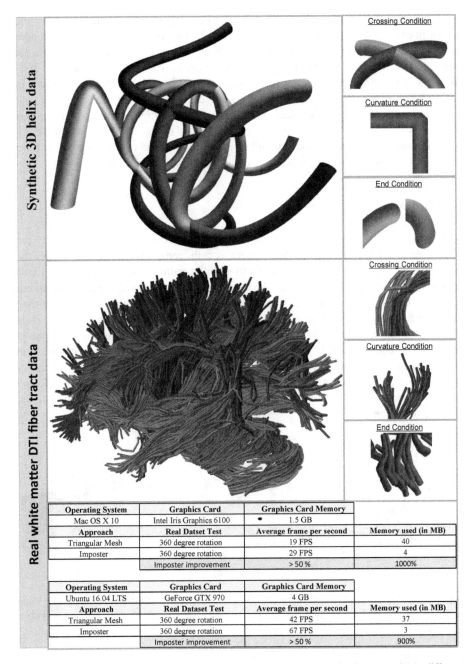

The table content within the figure:

Operating System	Graphics Card	Graphics Card Memory	
Mac OS X 10	Intel Iris Graphics 6100	1.5 GB	
Approach	**Real Datset Test**	**Average frame per second**	**Memory used (in MB)**
Triangular Mesh	360 degree rotation	19 FPS	40
Imposter	360 degree rotation	29 FPS	4
	Imposter improvement	> 50 %	1000%

Operating System	Graphics Card	Graphics Card Memory	
Ubuntu 16.04 LTS	GeForce GTX 970	4 GB	
Approach	**Real Dataset Test**	**Average frame per second**	**Memory used (in MB)**
Triangular Mesh	360 degree rotation	42 FPS	37
Imposter	360 degree rotation	67 FPS	3
	Imposter improvement	> 50 %	900%

Fig. 6. Visual experiments using the proposed imposter approach that test three different conditions: (1) Crossing, (2) Curvature, and (3) End. **Top:** 3D helix data set that has four different helix patterns. **Bottom:** Real white matter fiber tract data derived from a DTI scan. Included with real data is time and memory efficiency of imposter approach compared to fiber tracts rendered using cylinder triangular mesh. (Color figure online)

length of the smallest fiber tract is 3 points (2 line segments), and the length of the largest fiber tract is 39 points (38 line segments). In addition to the visual tests, render speed and memory efficiency performance tests are also provided on the real data set using the OpenGL implementation. More specifically, all 10,000 tracts are rotated 360° about the y-axis, and the average frame per second (fps) is recorded. As shown in Fig. 6, when compared to fiber tracts rendered using a cylinder triangle mesh, our imposter approach is over 50% faster on a GPU-based graphics cards, and requires 900% less memory. This is a significant savings in rendering speed and memory consumption, that can be utilized by GPU capable mobile and tablet devices that have less resources, but require the same processing and visual performance.

4   Limitations

As shown in Fig. 7(a) when two rectangular billboards cross each other at exactly the same depth, and the position of the camera is orthogonal to the viewing plane, the crossing artifact in Fig. 7(c) occurs. Even though the normal vectors reflect light like a cylinder, because the billboard is flat, the depth cannot be accurately determined. In our approach, the depth function uses a less than or equal pixel calculation, which is then passed to a linear blending function seen in Fig. 7(b) that adjusts the opacity value. It is important to note, the authors in [1] references a fragment-shader solution but never provides any details, an [2] explains the approach is limited to tuboids with constant radius, but never provide any experiments that show their overall approach works for all camera angles. From our point of view, depth artifacts are still an open issue, and we're working with NVIDIA community to develop a suitable solution that fits into the pipeline architecture.

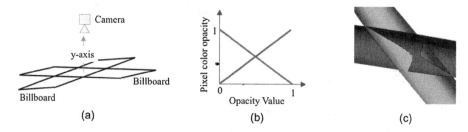

Fig. 7. Imposter limitation (a) the camera is directly above two billboards crossing at the same depth, (b) linear blending function, and (c) the visual result when depth and blending functions are applied.

5   Conclusion

We show that visualizing DTI fiber tracts using imposter approach decreases CPU/GPU computation complexity and memory requirements. The proposed approach is implemented using WebGL and OpenGL, and testing was performed on synthetic and real data sets. Lastly, the proposed imposter approach produces a suitable visual result for crossing, curvature, and end conditions common to white matter tractography data.

6   Implementation

The proposed imposter approach is implemented in the publically available SurfIce software application (https://www.nitrc.org/projects/surfice/). To enable fiber tract visualization using our imposter approach, open the SurfIce preferences dialog, then simply uncheck the better (but slower) tracks" check box illustrated in Fig. 8.

Fig. 8. How to enable fiber tract imposter approach in SurfIce Application

References

1. Schirski, M., et al.: Efficient visualization of large amounts of particle trajectories in virtual environments using virtual tubelets. In: Proceedings of the 2004 ACM SIGGRAPH International Conference on Virtual Reality Continuum and Its Applications in Industry, pp. 141–147. ACM, Singapore (2004)
2. Petrovic, V., Fallon, J., Kuester, F.: Visualizing whole-brain DTI tractography with GPU-based tuboids and LoD management. IEEE Trans. Vis. Comput. Graph. **13**(6), 1488–1495 (2007)
3. Angel, E., Shreiner, D.: Interactive Computer Graphics with WebGL. Addison-Wesley Professional, Upper Saddle River (2014)
4. Phong, B.T.: Illumination for computer generated pictures. Commun. ACM **18**(6), 311–317 (1975)

Revisiting Abnormalities in Brain Network Architecture Underlying Autism Using Topology-Inspired Statistical Inference

Sourabh Palande[1,2(✉)], Vipin Jose[1,2], Brandon Zielinski[3], Jeffrey Anderson[4], P. Thomas Fletcher[1,2], and Bei Wang[1,2]

[1] Scientific Computing and Imaging Institute,
University of Utah, Salt Lake City, USA
sourabh@sci.utah.edu
[2] School of Computing, University of Utah, Salt Lake City, USA
[3] Pediatrics and Neurology, University of Utah, Salt Lake City, USA
[4] Radiology, University of Utah, Salt Lake City, USA

Abstract. A large body of evidence relates autism with abnormal structural and functional brain connectivity. Structural covariance MRI (scMRI) is a technique that maps brain regions with covarying gray matter density across subjects. It provides a way to probe the anatomical structures underlying intrinsic connectivity networks (ICNs) through the analysis of the gray matter signal covariance. In this paper, we apply topological data analysis in conjunction with scMRI to explore network-specific differences in the gray matter structure in subjects with autism versus age-, gender- and IQ-matched controls. Specifically, we investigate topological differences in gray matter structures captured by structural covariance networks (SCNs) derived from three ICNs strongly implicated in autism, namely, the salience network (SN), the default mode network (DMN) and the executive control network (ECN). By combining topological data analysis with statistical inference, our results provide evidence of statistically significant network-specific structural abnormalities in autism, from SCNs derived from SN and ECN. These differences in brain architecture are consistent with direct structural analysis using scMRI (Zielinski et al. 2012).

1 Introduction

Autism is a complex developmental disorder characterized by impairment in social interactions, difficulty in verbal and nonverbal communications and repetitive behaviors. Although the exact mechanism of its development remains unclear, there is strong evidence relating autism to abnormal white matter and functional connectivity between brain regions. Structural abnormalities can be identified using voxel-based morphometry by comparing gray matter, white matter volumes, cortical thicknesses and their growth trajectories [11] across diagnostic groups. Although the gross brain differences have been well-documented [5], investigations into specific regional abnormalities in brain structure have

© Springer International Publishing AG 2017
G. Wu et al. (Eds.): CNI 2017, LNCS 10511, pp. 98–107, 2017.
DOI: 10.1007/978-3-319-67159-8_12

reported divergent results [14]. These inconsistent findings, however, may reflect discrete abnormalities in the brain network. Research has revealed a finite set of canonical domain-specific *resting state* or *intrinsic connectivity networks* (ICNs) that organize the brain function [8]. Many of the regions with reported abnormalities in autism lie within these ICNs. Network-specific differences could account for seemingly contradictory findings from previous studies.

Structural covariance MRI (scMRI) maps brain regions with covarying gray matter density across subjects, suggesting shared developmental or genetic influences. Seeley et al. [12] have used scMRI to demonstrate that specific brain disorders affect distinct ICNs and the corresponding gray matter regions. Using a similar technique, Zielinski et al. [16] have shown that there are network-specific structural differences between autism and control groups which are consistent with clinical aspects of the disease and that reported functional abnormalities in autism have a structural bias. Several recent studies have applied the scMRI technique to find network-specific structural abnormalities in other diseases such as Alzheimer's [10] and Huntington's [9].

scMRI identifies regions of gray matter that have a statistically significant correlation with a specific seed region of interest (ROI). We can model all pairwise correlations (across subjects) among the gray matter regions identified by the seed-based covariance map as a network. Comparing these networks across diagnostic groups may provide information not captured by direct comparisons between individual regions.

Several graph-theoretic measures have been proposed previously to compare networks [3]. However, a major drawback of these measures is their reliance on a fixed network topology. That is, these measures are typically based on a graph obtained by thresholding the connectivity matrix. The choice of threshold is crucial in such analyses. Different heuristics have been suggested to determine the threshold depending on which properties of the network are of interest. However, it is often not possible to determine a unique optimal threshold.

In this paper, we apply topological data analysis to structural covariance networks (SCNs) derived from three ICNs strongly implicated in autism; the default mode network (DMN), the salience network (SN) and the executive control network (ECN). Our method is based upon a core technique from topological data analysis known as *persistent homology* [6] where we extract topological features across all thresholds from a given network. We make use of topology-inspired statistical inference first reported by Chung et al. [4] to compare the extracted topological features. By combining topological data analysis with statistical inference, our results provide statistically significant evidence of structural abnormalities underlying SN and ECN in autism. Our results are consistent with the observations of Zielinski et al. [16] and may offer new insights towards interpreting fine-scale network-specific structural differences.

2   Technical Background

2.1   Structural Covariance Network

We use scMRI to extract a network-specific set of brain regions with covarying gray matter density across subjects. Given a seed ROI, separate condition-by-covariate analysis is performed for each gray matter region, in which the mean seed gray matter density is the covariate of interest and disease status is the grouping variable. Total brain volume (TBV) is included as a covariate-of-no-interest. This design enables us to determine the whole-brain patterns of seed-based structural covariance in each group. One-sample t-tests are performed to identify regions with significant groupwise gray matter density covariance with the seed ROI across subjects.

All pairwise correlations between gray matter densities across subjects, for pairs of identified regions, are modeled as a network. In what follows, we refer to such a network as the structural covariance network (SCN). The SCN, therefore, is a weighted, undirected graph $G(V, E, W)$, with gray matter regions as vertices and absolute values of pairwise correlations as edge weights. In particular, we compare SCNs generated with seed ROIs anchoring the three ICNs strongly implicated in autism, the SN, the ECN and the DMN. In the context of this paper, for simplicity (unless otherwise specified), we describe these SCNs by the name of their corresponding ICNs, namely, SN-SCN, ECN-SCN and DMN-SCN.

2.2   Graph Filtration

We extract topological features at multiple scales from a structural covariance network G by applying topological data analysis to a nested sequence of graphs constructed from G, referred to as the graph filtration.

Let $V = \{v_i \mid i = 1, \ldots, n\}$ be the vertex set with n vertices. Let E denote the edge set and W denote the set of edge weights. The edge between vertices v_i, v_j is denoted by e_{ij} and its weight is denoted by w_{ij}. $|E|$ denotes the number of edges. For a given threshold λ, we obtain a binary graph G_λ by removing edges with weight $w_{ij} \leq \lambda$. The adjacency matrix $A_\lambda = (a_{ij}(\lambda))$ is given by:

$$a_{ij}(\lambda) = \begin{cases} 0 & w_{ij} \leq \lambda \\ 1 & o.w. \end{cases}$$

As λ increases, more and more edges are removed from the graph. We can generate a sequence of thresholds, $\lambda_0 = 0 \leq \lambda_1 \leq \lambda_2 \leq \cdots \leq \lambda_q$, where $q \leq |E|$ by setting λ_i's equal to edge weights arranged in ascending order.

Corresponding to the sequence of thresholds we get a nested sequence of binary graphs, referred to as a *graph filtration* **G**:

$$G_{\lambda_0} \supseteq G_{\lambda_1} \supseteq G_{\lambda_2} \supseteq \cdots \supseteq G_{\lambda_q}.$$

We can measure the connectivity of a graph by its 0-th Betti number, β_0, which is the number of connected components in the graph. As the threshold λ increases,

$\beta_0(G_\lambda)$ of the corresponding graph also increases. The $\beta_0(G_{\lambda_i})$ of the graphs in filtration **G** form a monotonic sequence of integers,

$$\beta_0(G_{\lambda_0}) \le \beta_0(G_{\lambda_1}) \le \beta_0(G_{\lambda_2}) \le \cdots \le \beta_0(G_{\lambda_q}).$$

Suppose we start with a connected graph $G = G_{\lambda_0}$. We have $\beta_0(G_{\lambda_0}) = 1$ and $\beta_0(G_{\lambda_q}) = |V| = n$ by construction. Given n nodes, there are at most $\binom{2n}{n}$ unique edge weights; therefore $q \le \binom{2n}{n}$. The number of all possible monotonic integer sequences of length q, starting with 1 and ending with n, is finite.

Following the formulation of Chung et al. [4], the distance between two given graph filtrations **G** and **H** can be defined as:

$$D_q(\mathbf{G}, \mathbf{H}) = \sup_{0 \le i \le q} |\beta_0(G_{\lambda_i}) - \beta_0(H_{\lambda_i})|. \tag{1}$$

Intuitively, if we plot the two sequences of Betti numbers as a function of λ (the graph of such a function is referred to as the β_0 curve), this distance D_q measures the largest gap between the two curves. Given that the number of possible sequences is finite, D_q can take only a finite number of discrete integer values. Computing the β_0 curve for a given graph filtration could follow the standard algorithm for persistent homology [6]; in practice, a simpler algorithm can be used to capture the λ values when the number of components (clusters) decreases during the filtration.

2.3   Statistical Inference

We model the structural covariance networks for autism and control groups as weighted graphs G and H, respectively, with the corresponding graph filtrations **G** (autism) and **H** (control). We would like to test the equivalence of the two filtrations. In particular, we would like to test the null hypothesis H_0 against the alternative hypothesis H_1, where

$$H_0 : \beta_0(G_{\lambda_i}) = \beta_0(H_{\lambda_i}) \quad \text{for all } \lambda_i;$$

$$H_1 : \beta_0(G_{\lambda_i}) \ne \beta_0(H_{\lambda_i}) \quad \text{for some } \lambda_i.$$

By taking the *supremum* over all λ_i, D_q takes care of multiple comparisons implied in the hypotheses. Chung et al. [4] have provided a combinatorial derivation of the exact probability distribution of D_q. The proof is based on the Kolmogorov-Smirnov test [2]. This eliminates the need for numerically permuting samples for the test of hypothesis. The asymptotic probability distribution of D_q is given by:

$$\lim_{q \to \infty} P(D_q/\sqrt{2q} \ge d) = 2 \sum_{i=1}^{\infty} (-1)^{i-1} e^{-2i^2 d^2},$$

and the p-value under the null hypothesis can be computed as:

$$p = 2e^{-d_0^2} - 2e^{-8d_0^2} + 2e^{-18d_0^2} + \cdots \approx 2e^{-d_0^2} - 2e^{-8d_0^2} + 2e^{-18d_0^2},$$

where d_0 is the smallest integer greater than $D_q/\sqrt{2q}$.

3 Methods

3.1 Data Preprocessing

We derive our SCNs from the ICNs previously reported by Zielinski et al. [16, 17]. Here, we review the preprocessing pipeline. 49 male subjects with autism, aged 3–22 years, are compared to 49 age-, gender- and IQ-matched typically developing control subjects. Images are acquired using a Siemens 3.0 Tesla Trio MRI scanner. Whole brain isotropic MPRAGE image volumes are acquired in the sagittal plane using an 8-channel receive-only RF head coil, employing standard techniques (TR = 2300 ms, TE median = 3 ms, matrix median = $256 \times 256 \times 160$, flip angle = 12°, voxel resolution = 1 mm³, acquisition time = 9 min 12 s).

Customized image analysis templates are created by normalizing, segment- ing and averaging T1 images using SPM5 according to the processing pipeline proposed in [1,15]. First, images are transformed into standard space using a 12-parameter affine-only linear transformation and segmented into three tissue classes representing gray matter, white matter and cerebrospinal fluid. Then smoothly varying intensity changes as well as artifactual intensity alterations as a result of the normalization step are corrected for using a standard modulation algorithm within SPM5. Finally, the resulting segmented maps are smoothed using a 12-mm full-width at half-maximum Gaussian kernel.

In performing the scMRI analysis, a two-pass procedure is utilized, wherein study-specific templates are first created by segmenting our sample using a canonical pediatric template. Then tissue-specific prior probability maps are cre- ated from our sample. The tissue compartments are then re-segmented using this sample-specific template, so that the age range of our sample precisely matches that of the template(s) upon which the ultimate segmentations are based.

3.2 Structural Covariance Networks and Statistical Inference

The preprocessed images contain 7266 gray matter voxels. For each diagnostic group, a *whole-brain* SCN is constructed by computing pairwise correlations among all voxels.

To study network-specific structural covariance, 4-mm-radius spherical seed ROIs are selected within the right frontoinsular cortex (R FI) [12], the right dor- solateral prefrontal cortex (R DLPC) [13] and the right posterior cingulate cortex (R PCC) [7]. These regions anchor the salience network (SN), the executive con- trol network (ECN) and the default-mode network (DMN), respectively [7,12].

For each diagnostic group and each seed ROI, we generate SCNs following the process described in Sect. 2. The structural covariance maps corresponding to the seed ROI are shown in Fig. 1(a)-(c). The SCNs are composed of 4-mm-radius spherical regions identified by these maps. Further comparisons in Fig. 2 show that the two maps overlap in very few regions. Some regions present in the map for the control group are absent in the map for the autism group. Conversely, some regions are present only in the map for the autism group but not in the map for the control group. Figure 1(d) lists the number of regions present in controls but not in autism, in autism but not in controls and in both as well as in either autism or control.

We then construct and compare the SCNs derived from corresponding subsets of regions for each seed ROI. For each comparison, SCNs are derived for both diagnostic groups (autism and controls) separately. Graph filtrations are constructed for both networks. The distance D_q between the two resulting β_0 curves and the corresponding p-value for the test hypotheses is obtained accordingly.

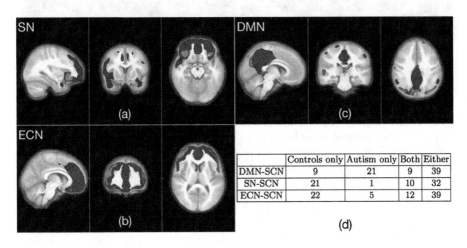

	Controls only	Autism only	Both	Either
DMN-SCN	9	21	9	39
SN-SCN	21	1	10	32
ECN-SCN	22	5	12	39

Fig. 1. (a)-(c) Structural covariance maps with seed in R FI, R DLPC and R PCC, anchoring SN, ECN and DMN, respectively. Red represents the autism group map, blue represents the control group map. (d) Number of regions identified from scMRI map for a given seed region. (Color figure online)

4   Results

We apply statistical inference and compare SCNs across groups of subjects with autism and typically developing control subjects. We begin by comparing the global SCNs composed of all 7266 gray matter voxels in the preprocessed images. Applying the statistical inference detailed in Sect. 3, we obtain a p-value of $6.6250179 \times 10^{-19}$. The differences in whole-brain gray matter composition between the autism and control groups have been well established in previous studies [5]. The near-zero p-value shows that such differences can also be captured by the topological features extracted from the global SCNs.

For a closer analysis, we compare the SCNs generated with seed ROIs anchoring the three ICNs (SN, ECN and DMN), referred to as SN-SCN, ECN-SCN and DMN-SCN, respectively. Recall that the structural covariance maps for the autism and the control groups overlap in very few regions. We construct and compare SCNs derived from subsets of regions that are present in controls but not in autism, present in autism but not in controls and present in both as well as present in either.

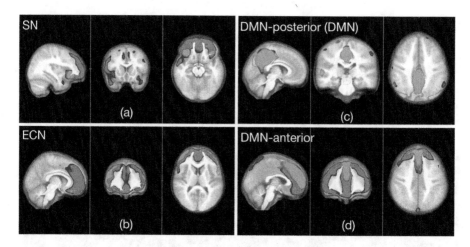

Fig. 2. scMRI maps are further illustrated here with red to yellow (autism) and dark blue to light blue (control) color look up tables. The color gradation indicates increasing statistical significance. The overlapping regions among the autism and control groups are highlighted in green. Note for (c) and (d): Our data consists of subjects with an average age of about 13 years. The underlying structure of the DMN is not fully developed at this age. We include two DMN maps with different seeds to show that the posterior part (c) is not yet integrated with the anterior part (d). In our analysis, we use the posterior covariance map (c) which corresponds to the most common seed for DMN in adults (R PCC). (Color figure online)

The β_0 curves corresponding to comparisons among global SCNs, and seed-specific SCNs generated from regions present either in autism or controls, are shown in Fig. 3. Table 1 lists the p-values obtained by applying the statistical inference procedure to the corresponding SCNs. By combining topological data analysis with statistical inference, our results provide statistically significant evidence of network-specific structural abnormalities in autism for both SN-SCNs and ECN-SCNs.

Table 1. p-values for statistical inference on SCNs derived from ICNs; DMN-SCNs, SN-SCNs and ECN-SCNs. Only one region in SN is present in autism but not in controls where the inference procedure is not applicable.

	Controls only	Autism only	Both	Either
DMN-SCN	0.6271670	0.0815188	0.9538228	0.2369032
SN-SCN	**0.0014932**	NA	**0.0366311**	**1.3269078×10^{-6}**
ECN-SCN	**0.0422562**	0.9960098	**0.0059460**	**1.7996732×10^{-6}**

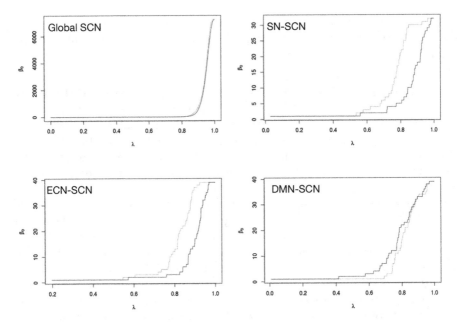

Fig. 3. β_0 curves from Global SCNs as well as SN-SCNs, ECN-SCNs and DMN-SCNs, generated from regions present in either autism (red) or controls (green) respectively. (Color figure online)

5   Conclusion and Discussion

Using direct comparisons of structural covariance maps, Zielinski et al. have shown the structural differences in gray matter regions underlying *intrinsic connectivity networks* (ICNs) such as SN [16], DMN [16] and ECN (Brandon Zielinski, personal communication, May 2017) between the autism and the control groups. In contrast, our method compares the *structural covariance networks* (SCNs), which are composed of all possible pairwise correlations between gray matter regions and not just their covariance with a specific seed region.

Our inference procedure obtains statistically significant *p*-values among the SCNs derived from SN and ECN (SN-SCNs and ECN-SCNs) when comparing networks constructed from regions present in controls only, regions present in both autism and controls, as well as regions present in either autism or controls. Our results indicate statistically significant differences in the 0-dimensional topological features of these SCNs; this result is consistent with the findings of Zielinski et al. [16].

Our method, however, does not capture significant differences in the topology of SCNs derived from DMN (DMN-SCNs). It is possible that considering only pairwise interactions among gray matter regions (that is, 0-order topological features encoded by the β_0 curves, corresponding to the number of connected components) may not be sufficient to capture the complex topological differences within these SCNs. Analyzing three-way or four-way interactions, capturing higher-order topological features such as tunnels and voids and focusing on specific sites directly involved in merging components in the graph filtration may provide further insights into these SCNs.

Final Remarks. A key insight from the work of Zielinski et al. [16] is that structure enables function and functional collaboration enables structure. Our work, in particular, helps summarize these multidimensional, structure-function relationships by conceptualizing them as higher order topological relationships. The SCNs are constructed using inputs from both the function, in form of the seed ROI anchoring specific ICNs and the structure, in the form of gray matter density signals.

The techniques in [16] compare covariance maps directly. Such a comparison helps to identify whether a particular region is present or absent in the autism vs the control group maps. The regions in these maps are assigned significance measures using their covariance with respect to a specific seed region. Our work, on the other hand, uses the SCNs to encode all pairwise associations among regions, where the extent of an association is measured by the correlations across subjects. Our results indicate that there are statistically significant differences in the way networks are connected, which implies differences in the patterns of pairwise association across diagnostic groups.

To illustrate the advantage, consider the regions present in the SN- or ECN-specific covariance maps of both the autism and the control groups. Direct comparison of the covariance maps does not provide any further insight into these regions. Our method on the other hand, shows that there are statistically significant differences in the topological features derived from SN-SCN and ECN-SCN composed of these regions (Table 1, p-values of 0.0366311 and 0.0059460 respectively). However, it should be noted that the SCNs are abstract networks and do not represent physical connectivity between the regions. This limits the interpretability of our results to some extent. Further analysis is needed in order to quantify and better interpret the differences in the SCNs suggested by the statistical inference.

Acknowledgements. This work was supported by NIH grant R01EB022876 and NSF grant IIS-1513616.

References

1. Altaye, M., Holland, S.K., Wilke, M., Gaser, C.: Infant brain probability templates for MRI segmentation and normalization. NeuroImage **43**(4), 721–730 (2008)
2. Böhm, W., Hornik, K.: A Kolmogorov-Smirnov test for r samples. Fundam. Inf. **117**(1–4), 103–125 (2012)
3. Bullmore, E., Sporns, O.: Complex brain networks: graph theoretical analysis of structural and functional systems. Nat. Rev. Neurosci. **10**(3), 186–198 (2009)
4. Chung, M.K., Villalta-Gil, V., Lee, H., Rathouz, P.J., Lahey, B.B., Zald, D.H.: Exact topological inference for paired brain networks via persistent homology. bioRxiv:140533 (2017)

5. Courchesne, E., Pierce, K., Schumann, C.M., Redcay, E., Buckwalter, J.A., Kennedy, D.P., Morgan, J.: Mapping early brain development in autism. Neuron **56**(2), 399–413 (2007)
6. Edelsbrunner, H., Letscher, D., Zomorodian, A.J.: Topological persistence and simplification. Discrete Comput. Geom. **28**, 511–533 (2002)
7. Fair, D.A., Cohen, A.L., Dosenbach, N.U.F., Church, J.A., Miezin, F.M., Barch, D.M., Raichle, M.E., Petersen, S.E., Schlaggar, B.L.: The maturing architecture of the brain's default network. Proc. Natl. Acad. Sci. **105**(10), 4028–4032 (2008)
8. Fox, M.D., Snyder, A.Z., Vincent, J.L., Corbetta, M., Van Essen, D.C., Raichle, M.E.: The human brain is intrinsically organized into dynamic, anticorrelated functional networks. Proc. Natl. Acad. Sci. U.S.A. **102**(27), 9673–9678 (2005)
9. Minkova, L., Eickhoff, S.B., Abdulkadir, A., Kaller, C.P., Peter, J., Scheller, E., Lahr, J., Roos, R.A., Durr, A., Leavitt, B.R., Tabrizi, S.J., Klöppel, S., Investigators, T.-H.: Large-scale brain network abnormalities in huntington's disease revealed by structural covariance. Hum. Brain Mapp. **37**(1), 67–80 (2016)
10. Montembeault, M., Rouleau, I., Provost, J.-S., Brambati, S.M.: Altered gray matter structural covariance networks in early stages of Alzheimer's disease. Cereb. Cortex **26**(6), 2650 (2016)
11. Schumann, C.M., Bloss, C.S., Barnes, C.C., Wideman, G.M., Carper, R.A., Akshoomoff, N., Pierce, K., Hagler, D., Schork, N., Lord, C., Courchesne, E.: Longitudinal magnetic resonance imaging study of cortical development through early childhood in autism. J. Neurosci. **30**(12), 4419–4427 (2010)
12. Seeley, W.W., Crawford, R.K., Zhou, J., Miller, B.L., Greicius, M.D.: Neurodegenerative diseases target large-scale human brain networks. Neuron **62**, 42–52 (2009)
13. Seeley, W.W., Menon, V., Schatzberg, A.F., Keller, J., Glover, G.H., Kenna, H., Reiss, A.L., Greicius, M.D.: Dissociable intrinsic connectivity networks for salience processing and executive control. J. Neurosci. **27**(9), 2349–2356 (2007)
14. Stigler, K.A., McDonald, B.C., Anand, A., Saykin, A.J., McDougle, C.J.: Structural and functional magnetic resonance imaging of autism spectrum disorders. Brain Res. **1380**, 146–161 (2011)
15. Wilke, M., Holland, S.K., Altaye, M., Gaser, C.: Template-O-Matic: a toolbox for creating customized pediatric templates. NeuroImage **41**(3), 903–913 (2008)
16. Zielinski, B.A., Anderson, J.S., Froehlich, A.L., Prigge, M.B.D., Nielsen, J.A., Cooperrider, J.R., Cariello, A.N., Fletcher, P.T., Alexander, A.L., Lange, N., Bigler, E.D., Lainhart, J.E.: scMRI reveals large-scale brain network abnormalities in Autism. PLOS ONE **7**(11), 1–14 (2012)
17. Zielinski, B.A., Gennatas, E.D., Zhou, J., Seeley, W.W.: Network-level structural covariance in the developing brain. Proc. Natl. Acad. Sci. **107**(42), 18191–18196 (2010)

"Evaluating Acquisition Time of rfMRI in the Human Connectome Project for Early Psychosis. How Much Is Enough?"

Sylvain Bouix[1(✉)], Sophia Swago[1], John D. West[2], Ofer Pasternak[1,3],
Alan Breier[4], and Martha E. Shenton[1,3,5]

[1] Psychiatry Neuroimaging Laboratory, Department of Psychiatry, Harvard Medical School,
Brigham and Women's Hospital, Boston, MA, USA
sylvain@bwh.harvard.edu
[2] Department of Radiology and Imaging Sciences, Center for Neuroimaging,
Indiana University School of Medicine, Indianapolis, IN, USA
[3] Department of Radiology, Harvard Medical School, Brigham and Women's Hospital,
Boston, MA, USA
[4] Department of Psychiatry, Indiana University School of Medicine, Indianapolis, IN, USA
[5] Veteran Affairs Boston Healthcare System, Brockton Division, Harvard Medical School,
Brockton, MA, USA

Abstract. Resting-state functional MRI (rfMRI) correlates activity across brain regions to identify functional connectivity networks. The Human Connectome Project (HCP) for Early Psychosis has adopted the protocol of the HCP Lifespan Project, which collects 20 min of rfMRI data. However, because it is difficult for psychotic patients to remain in the scanner for long durations, we investigate here the reliability of collecting less than 20 min of rfMRI data. Varying durations of data were taken from the full datasets of 11 subjects. Correlation matrices derived from varying amounts of data were compared using the Bhattacharyya distance, and the reliability of functional network ranks was assessed using the Friedman test. We found that correlation matrix reliability improves steeply with longer windows of data up to 11–12 min, and ≥14 min of data produces correlation matrices within the variability of those produced by 18 min of data. The reliability of network connectivity rank increases with increasing durations of data, and qualitatively similar connectivity ranks for ≥10 min of data indicates that 10 min of data can still capture robust information about network connectivities.

Keywords: Resting state · Acquisition time · Bhattacharyya distance

1 Introduction

Resting-state functional magnetic resonance imaging (rfMRI) can be used to correlate spontaneous fluctuations in the blood oxygen level-dependent (BOLD) signal across regions of the resting brain in order to identify functional connectivity networks [1]. rfMRI is well suited to studying patient populations compared to task-based functional MRI [2] because it does not require participants to perform any tasks, reducing patient

© Springer International Publishing AG 2017
G. Wu et al. (Eds.): CNI 2017, LNCS 10511, pp. 108–115, 2017.
DOI: 10.1007/978-3-319-67159-8_13

burden and expanding the patient pool to those whose cognitive or physical impairment would otherwise preclude them, and has good signal to noise ratio. rfMRI has potential applications in studying group differences, evaluating treatment, and as a diagnostic biomarker [2]. In recent years, it has increasingly been used to investigate changes in functional networks arising from neurological and psychiatric disorders, including stroke, Alzheimer's disease, depression, and schizophrenia [3].

An open question in rfMRI methodology is how much data to acquire. Typical rfMRI studies only acquire 5–10 min of data, but the reliability of estimating functional networks improves with increased scan time. However, there are varying recommendations for how many minutes of data are sufficient. While Van Dijk et al. [4] reported that connectivity estimates stabilized at ~5 min of rfMRI data, Anderson et al. [5] suggested at least 25 min be collected, and recently, Laumann et al. [6] saw further gains in reliability with up to 100 min of data. The Washington University-University of Minnesota Human Connectome Project (HCP) [7] collected 60 min of rfMRI in four runs of 15 min. The relatively long scan time was intended to counteract SNR loss due to the high spatial resolution of the data [8].

Our study collects Connectome-like high quality imaging data in an early psychosis population. We have adopted the protocol of the shorter HCP Pilot Lifespan Project [9], which collects ~60 min total of imaging data and includes ~20 min of rfMRI data, to make the scan time more tolerable for patients. However, because it is difficult for psychotic patients to remain in the scanner for long durations, we here investigate the reliability of connectivity estimates from less than 20 min of rfMRI data.

2   Methods

2.1   Participants

Eleven healthy male participants between the ages of 16 and 35 were recruited from Indiana University, Beth Israel Deaconess Medical Center, Massachusetts General Hospital, and McLean Hospital. The Structured Clinical Interview for DSM-5-RV (SCID) [10] was administered to confirm that subjects did not meet criteria for psychiatric diagnosis. All subjects gave written informed consent.

2.2   Data Acquisition

Imaging data were collected on two Siemens Prisma 3.0 T MRI scanners at Indiana University and Brigham and Women's Hospital with a 32-channel head coil using a modified protocol based on the HCP LifeSpan Pilot project at the University of Minnesota [9]. Structural, functional, and diffusion data were acquired, with a total scan time of about 70 min. Structural data included a T1-weighted image (MPRAGE, 0.8 mm^3 isotropic resolution, TR/TE = 2400/2.22 ms, TI = 1000 ms, flip-angle = 8°) and a T2-weighted image (SPACE, 0.8 mm^3 isotropic resolution, TR/TE = 3200/563 ms). Multi-shell diffusion data were acquired four times, twice with anterior-posterior (A-P) phase encoding and twice with posterior-anterior (P-A) phase encoding (1.5 mm^3 isotropic resolution, 92 directions, TR/TE = 3230/89.2 ms, multiband acceleration factor = 4, b = 1500 and 3000). rfMRI data were also acquired four times, twice with A-P phase encoding and twice with P-A phase encoding (EPI, 2.0 mm^3 isotropic resolution, TR/ TE = 800/37 ms, multiband acceleration factor = 8). Each functional run consisted of 420 time points, totaling 5.6 min. Subjects were instructed to keep their eyes open during all functional scans. In addition, spin echo field maps were acquired to correct for intensity and geometric distortions.

2.3   Data Preprocessing

Structural and functional data were preprocessed with the HCP Minimal Preprocessing Pipelines [11]. The structural pipeline entails registration of the T1w and T2w images, bias field correction, registration to a MNI space template to allow comparison across subjects, reconstruction of white and pial surfaces, and surface registration to a surface atlas. Finally, the data are converted to the Connectivity Informatics Technology Initiative (CIFTI) file format, which combines surface cortical data and volumetric subcortical data in a single "grayordinate" coordinate system.

rfMRI preprocessing includes echo planar imaging (EPI) distortion correction, rfMRI to T1w image registration, motion correction, intensity normalization, and conversion to CIFTI format. The rfMRI data were additionally processed with the Oxford University Centre for Functional MRI of the Brain (FMRIB) group's independent component analysis-based Xnoiseifer - FIX (ICA FIX) [12] pipeline to decompose the BOLD signal and regress out components of physiological noise and noise caused by motion. The resulting time series were then de-meaned and normalized by the standard deviation [13] and concatenated to produce a single time series of 1,680 time points (22.4 min).

Contiguous windows of varying durations, ranging from 5.6 to 18 min, were randomly extracted from the full 22.4 min dataset. Durations above 18 min were not extracted because the number of different windows that could be obtained would be limited. One hundred windows were extracted for each duration (i.e. 100 of 5.6 min, 100 of 6 min, etc.). Three *a priori* parcellation schemes (Fig. 1) were then applied to each window of data: the HCP Nature parcellation (360 regions) [14], Yeo's 17-network parcellation (100 regions), and Yeo's 7-network parcellation (8 regions) [15]. Using Pearson's correlation coefficient, region-to-region functional correlation matrices were generated for all parcellations of each window of data. Using the 17-network parcellation

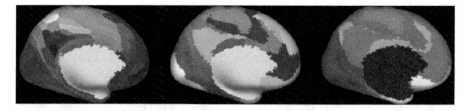

Fig. 1. Different rfMRI parcellation schemes of varying granularity shown on an inflated surface. Left: HCP Nature (360 regions), Center: Yeo 17-network (100 regions), and Right: Yeo 7-network (8 regions).

only, the mean within-network connectivities were calculated by averaging the correlations found between regions belonging to the same network. Each network was ranked from lowest to highest mean within-network connectivity.

2.4 Statistical Analysis

Correlation Matrix Reliability. Each duration of data was directly compared to each of the 18-minute windows within each parcellation to determine matrix reliability. Correlation matrices were directly compared using the Bhattacharyya distance D_B [16], which measures the similarity of two probability distributions. The distance D_B for two multivariate normal distributions with means (μ_1, μ_2) and covariance matrices (Σ_1, Σ_2) has the form

$$D_B = \frac{1}{8}\left(\mu_1 - \mu_2\right)^T \Sigma^{-1}\left(\mu_1 - \mu_2\right) + \frac{1}{2}\ln\left(\frac{|\Sigma|}{\sqrt{|\Sigma_1||\Sigma_2|}}\right), \tag{1}$$

where $\Sigma = (\Sigma_1 + \Sigma_2)/2$ and $|\ |$ signifies the matrix determinant. In the current study, Σ_2 was always a correlation matrix derived from an 18-minute window. Because the time series have equal means as a result of de-meaning during preprocessing, Eq. 1 becomes

$$D_B = \frac{1}{2}\ln\left(\frac{|\Sigma|}{\sqrt{|\Sigma_1||\Sigma_2|}}\right). \tag{2}$$

As Σ_1 and Σ_2 become more similar, D_B approaches zero.

Network Connectivity Rank Reliability. Networks using the 17-Network parcellation were ranked from lowest to highest within-network connectivity. Changes in network ranks across increasing durations of data were evaluated using the Friedman test, a nonparametric repeated measures analysis, and results were corrected for multiple comparisons.

3 Results

3.1 Correlation Matrix Reliability

A steep decrease in D_B, indicating increasingly similar correlation matrices, was observed up to the maximal curvature point observed at durations of 11.21 ± 0.74, 11.97 ± 0.89, and 12.38 ± 1.26 min for the HCP Nature, Yeo 17-network, and Yeo-7 network parcellations respectively (Fig. 2). Note that as the number of regions in the parcellation decreases, the improvement in matrix similarity is less dramatic. D_B then more slowly approached zero. The middle 95% range of D_B values comparing 18-minute windows against other 18-minute windows for each parcellation represents the variability of correlation matrices at the maximum duration. The upper limit of this range

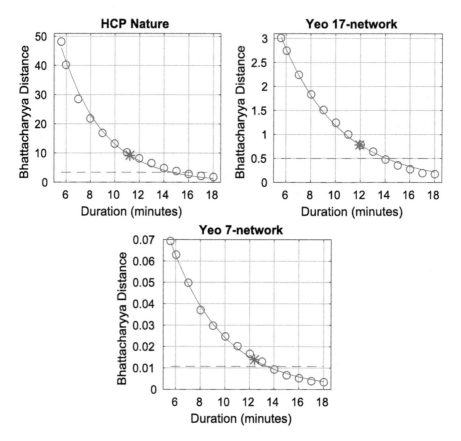

Fig. 2. Mean D_B for each duration compared to 18 min of data from three parcellations of a representative subject. Points of maximum curvature (star) and middle 95% ranges of 18 min vs 18 min D_B values (dotted lines) are shown.

was taken as a threshold to determine at which duration correlation matrices became comparable to a matrix derived from 18 min. The average durations where D_B began to fall within the variability of 18-minute vs 18-minute D_B values were 14.63 ± 1.32, 14.08 ± 1.58, and 13.43 ± 1.97 min for the HCP Nature, Yeo 17-network, and Yeo-7 network parcellations respectively (Fig. 2).

3.2   Network Rank Reliability

The Friedman test was used to test for a significant difference in mean within-network connectivity rank across varying durations of data. 93.58 ± 6.68% of networks had at least one duration that significantly differed in rank ($p < 0.029$) from of 5.6 to 18 min of data. This percentage decreased to 79.14 ± 6.64% for rank changes over the range of 10 to 18 min, and further to 62.57 ± 8.43% from 14 to 18 min.

Networks with high within-network connectivity (ranks 16 and 17) were highly reliable even at durations as short as 5–6 min (Fig. 4). These ranks primarily corresponded to networks 1 and 2, the peripheral and central visual networks. This agrees with Gonzalez-Castillo, et al. [17] who found the visual network as one of the most temporally stable networks for within-network connections. Many other networks began to display stable rankings at 10–12 min of data.

4   Discussion and Conclusion

Using the Bhattacharyya distance, we found that the reliability of functional correlation matrices derived from rfMRI improves monotonically as more data is acquired. The improvement reaches an inflexion point with windows of data around 11–12 min, at which point acquiring more data is less beneficial. Fourteen minutes or more of data was found to produce correlation matrices within the variability of those produced by 18 min of data. Since all D_B values were calculated as a comparison to 18-minutes of data, these findings are specific to a study whose gold standard is ~ 20 min of rfMRI data. Nonetheless, our method of using the Bhattacharyya distance can be used by other studies to determine how much data is necessary for a session to be usable in further analyses. This is especially valuable for populations that may not tolerate the full scanning protocol of the study.

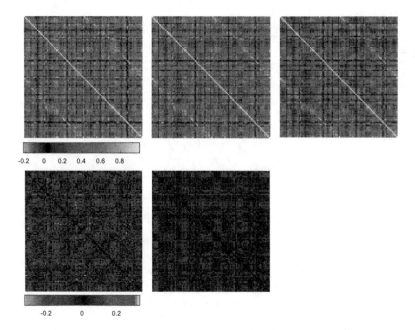

Fig. 3. Top: Representative connectivity matrices derived from 5 min (left), 11 min (center), and 18 min (right) of data using the Yeo 17-network parcellation from a single subject. Bottom: Connectivity differences between 18 and 5 min of data (left) and between 18 and 11 min of data (right).

Our results show that within-network connectivity rank reliability also increases with increasing durations of data, although our statistical tests are not ideally suited for rankings with a zero variance (e.g. a network is **always** ranked the same in all experiments for a particular duration, as network 4 was for all durations except 5.6 min (Fig. 4)). Indeed, we found that many networks qualitatively exhibited similar connectivity ranks starting at 10–12 min of data, indicating 10–12 min of data may be a suitable amount for many network connectivity analyses. Our results also indicate that the duration required is dependent on the network of interest, as some networks, such as the central and peripheral visual network, are very reliable with as little as 5–6 min of data.

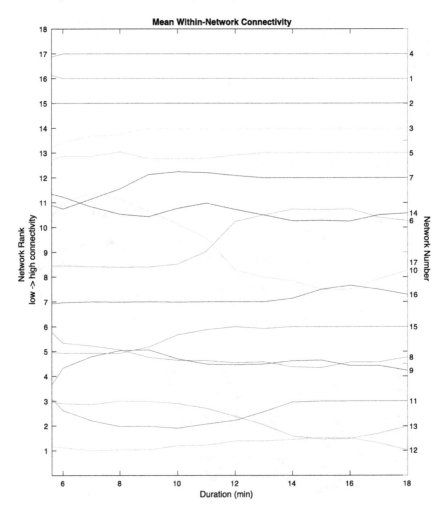

Fig. 4. Mean within-network connectivity ranks from a representative subject using 100 randomly sampled windows of data for each duration.

Acknowledgments. This research was supported by National Institutes of Health (NIH) grants U01MH109977, R01MH085953, and P41EB015902.

References

1. Biswal, B., Yetkin, F.Z., Haughton, V.M., Hyde, J.S.: Functional connectivity in the motor cortex of resting human brain using echo-planar MRI. Magn. Reson. Med. **34**(4), 537–541 (1995)
2. Fox, M.D., Greicius, M.: Clinical applications of resting state functional connectivity. Front Syst. Neurosci. **4**, 19 (2010). doi:10.3389/fnsys.2010.00019
3. Snyder, A.Z., Raichle, M.E.: A brief history of the resting state: the Washington University perspective. NeuroImage **62**(2), 902–910 (2012). doi:10.1016/j.neuroimage.2012.01.044
4. Van Dijk, K.R.A., Hedden, T., Venkataraman, A., Evans, K.C., Lazar, S.W., Buckner, R.L.: Intrinsic functional connectivity as a tool for human connectomics: theory, properties, and optimization. J. Neurophysiol. **103**(1), 297–321 (2010). doi:10.1152/jn.00783.2009
5. Anderson, J.S., Ferguson, M.A., Lopez-Larson, M., Yurgelun-Todd, D.: Reproducibility of single-subject functional connectivity measurements. AJNR Am. J. Neuroradiol. **32**(3), 548–555 (2011). doi:10.3174/ajnr.A2330
6. Laumann, T.O., Gordon, E.M., Adeyemo, B., et al.: Functional system and areal organization of a highly sampled individual human brain. Neuron **87**(3), 657–670 (2015). doi:10.1016/j.neuron.2015.06.037
7. Van Essen, D.C., Smith, S.M., Barch, D.M., et al.: The WU-Minn Human Connectome Project: An Overview. NeuroImage **80**, 62–79 (2013). doi:10.1016/j.neuroimage.2013.05.041
8. Smith, S.M., Andersson, J., Auerbach, E.J., et al.: Resting-state fMRI in the Human Connectome Project. NeuroImage **80**, 144–168 (2013). doi:10.1016/j.neuroimage.2013.05.039
9. Phase 1b Lifespan Pilot Parameters, http://lifespan.humanconnectome.org/data/phase1b-pilot-parameters.html. Accessed 09 Jun 2017
10. First, M.B., Williams, J.B.W., Karg, R.S., Spitzer, R.L.: Structured Clinical Interview for DSM-5—Research Version (SCID-5 for DSM-5, Research Version; SCID-5-RV). American Psychiatric Association, Arlington (2015)
11. Glasser, M.F., Sotiropoulos, S.N., Wilson, J.A., et al.: The minimal preprocessing pipelines for the human connectome project. NeuroImage **80**, 105–124 (2013). doi:10.1016/j.neuroimage.2013.04.127
12. Salimi-Khorshidi, G., Douaud, G., Beckmann, C.F., Glasser, M.F., Griffanti, L., Smith, S.M.: Automatic denoising of functional MRI data: combining independent component analysis and hierarchical fusion of classifiers. NeuroImage **90**, 449–468 (2014). doi:10.1016/j.neuroimage.2013.11.046
13. Beckmann, C.F., Smith, S.M.: Probabilistic independent component analysis for functional magnetic resonance imaging. IEEE Trans. Med. Imaging **23**(2), 137–152 (2004). doi:10.1109/TMI.2003.822821
14. Glasser, M.F., Coalson, T.S., Robinson, E.C., et al.: A multi-modal parcellation of human cerebral cortex. Nature **536**(7615), 171–178 (2016). doi:10.1038/nature18933
15. Yeo, B.T., Krienen, F.M., Sepulcre, J., et al.: The organization of the human cerebral cortex estimated by intrinsic functional connectivity. J. Neurophysiol. **106**(3), 1125–1165 (2011). doi:10.1152/jn.00338.2011
16. Bhattacharyya, A.: On a measure of divergence between two statistical populations defined by their probability distributions. Bull. Calcutta Math. Soc. **35**, 99–109 (1943)
17. Gonzalez-Castillo, J., Handwerker, D.A., Robinson, M.E., et al.: The spatial structure of resting state connectivity stability on the scale of minutes. Front Neurosci. **8**, 138 (2014). doi:10.3389/fnins.2014.00138

Early Brain Functional Segregation and Integration Predict Later Cognitive Performance

Han Zhang[1,2], Weiyan Yin[2,3], Weili Lin[1,2,3], and Dinggang Shen[1,2,4(✉)]

[1] Department of Radiology, University of North Carolina at Chapel Hill, Chapel Hill, NC, USA
dgshen@med.unc.edu
[2] BRIC, University of North Carolina at Chapel Hill, Chapel Hill, NC, USA
[3] Biomedical Engineering, University of North Carolina at Chapel Hill, Chapel Hill, NC, USA
[4] Department of Brain and Cognitive Engineering, Korea University, Seoul, Republic of Korea

Abstract. The human brain in the first 2 years of life is fascinating yet mysterious. Whether its connectivity pattern is genetically predefined for neonates and predictive to the later cognitive performance is unknown. Numerous neurological/psychiatric diseases in adults with impaired cognitive functions have been linked with deteriorated "triple networks" that govern the high-level cognition. The triple networks are referred to salience network for salient event monitoring and emotion processing, default mode network for self-cognition and episodic memory, and executive control network for attention control, set maintenance and task executions. We investigate the infancy "triple networks" and their development in the pivotal period of the first two years of life with longitudinal resting-state fMRI from 52 term infants (24 having cognitive performance scores tested at 4 years old). We found that the triple networks harbor at the medial prefrontal cortex, an ideal brain region for unveiling early development of the high-level functions. Further parcellation of this area indicates consistent subdivisions from 0 to 2 years old, indicating largely predefined functional segregation in this highly heterogeneous region. Interconnectivity among the mediofrontal subdivisions reveals a significant invert U-shape curve for modularity, with the inter-network functional connectivity (FC) peaking at 6–9 months, manifesting a developing functional integration within the frontal region. Through long-range FC, we found the development of the high-level functions starts from salience monitoring, followed by self-cognition, then to executive control. We extract both within-frontal modularity index (reflecting short-distance FC), and outreaching index (measuring long-distance FC) for the newborns. Interestingly, these connectomics features for the newborns well predict their later cognitive performance at 4 years old. These results converge to favoring a predefined genetic dominance in the development of triple networks' FC, which is essential for understanding early high-level neuro-cognitive development and promising for early abnormality detection.

Keywords: Functional connectivity · functional Magnetic Resonance Imaging (fMRI) · Resting-state fMRI · Brain networks · Infant · Development · Modularity

© Springer International Publishing AG 2017
G. Wu et al. (Eds.): CNI 2017, LNCS 10511, pp. 116–124, 2017.
DOI: 10.1007/978-3-319-67159-8_14

1   Introduction

Many psychiatric diseases and mental disorders occur at early adolescence [1]. The most relevant central nervous system is the "triple networks" [2] which constitute three high-level cognitive function related functional systems: salience network (SN) for salient event monitoring and emotion processing, default mode network (DMN) for self-cognition and episodic memory, and executive control network (ECN) for attention control, set maintenance and task executions. While adult triple networks are the researching hotspots, their infancy patterns and early developmental trajectories are, however, largely unknown. Based on previous studies [3, 4], the triple networks, as all other high-level cognitive function-related functional systems, are not well developed at birth; thus both genetic and environmental factors may affect their prolonged developmental curves and their dynamic interactions [5]. Understanding such a process is critical for early diagnosis of grievous later consequences [6].

In this work, based on longitudinal resting-state fMRI (rs-fMRI) from healthy term infants scanned at birth and every three months after birth, we developed a computational framework and a set of new quantification metrics and characterized the functional segregation and integration of such important triple networks in this pivotal early life period. For the first time, we revealed developing associations between the early brain functional connectivity (FC) architectures of the neonates and their later cognitive ability tested at four years old. We will show in next sections how to *(1)* build an early developing triple network model; *(2)* characterize its functional-anatomical profiles; *(3)* investigate the association between later cognitive ability and the earlier developing curve longitudinally; and *(4)* individually predict later cognitive ability based on the newborns' triple-network FC patterns. Although this is an early prediction study for healthy infants, it is straightforward to apply this framework to high-risk population screening and early diagnosis of neurodevelopmental diseases.

2   Methods and Results

2.1   An Early Developing Triple Network Model

The adult triple networks consist of the SN, DMN and ECN, which encompass a large portion of the brain. However, we found that, with an overlapping analysis, the triple networks highly overlap in the mediofrontal area that includes both dorsal anterior cingulate cortex and other midline structures in the prefrontal cortex. In other brain regions, the overlap is quite sparse (Fig. 1). This indicates that the mediofrontal area is functionally heterogeneous and densely interconnected. Therefore, this area becomes an ideal model for investigating early development of brain functional segregation and integration. For this initial first study of triple networks in longitudinal neonate data, we focus on the specific brain region where the three networks are known to overlap. Based on the previous adult studies, this area contains three major subdivisions: (1) The dorsomedial prefrontal part that connects to the ECN, (2) the ventromedial prefrontal subregion that belongs to the DMN, and (3) the dorsal anterior cingular gyrus where the SN anchors (Fig. 1). To answer the question that whether the mediofrontal area has already been functionally specific to the three networks during infancy, we first conduct functional segmentation in this area. We compare the result with that from adults to see if such a configuration presents after birth and how it evolves during the first 2-year development.

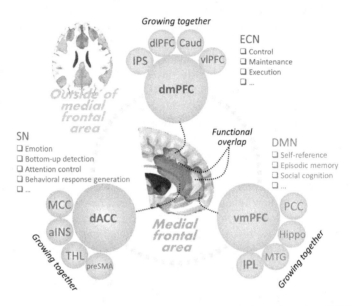

Fig. 1. The triple network model with the highlighted mediofrontal portion. ECN: executive control network; SN: salience network; DMN: default mode network; dmPFC: dorsomedial prefrontal cortex; dACC: dorsal anterior cingulate cortex; vmPFC: ventromedial PFC; IPS: intraparietal sulcus; dlPFC: dorsolateral PFC; Caud: caudate; vlPFC: ventrolateral PFC; MCC: middle cingulate cortex; aINC: anterior insula; THL: thalamus; preSMA: pre-supplementary motor area; IPL: inferior parietal lobule; MTG: middle temporal gyrus; Hippo: hippocampus; PCC: posterior cingulate cortex.

We use the longitudinal 5-min rs-fMRI data (150 volumes, repetition time = 2 s, voxel size = $4 \times 4 \times 4$ mm^3) of 52 infants, most of whom were scanned at birth and 3, 6, 9, 12, 18 and 24 months old. We have 33, 29, 31, 30, 35, 26 and 22 samples scanned at the above seven time points, respectively. Data preprocessing is performed with FMRIBs Software Libraries (FSL) according to previous studies [4], where image registration is particularly conducted with a 4-D group-wise longitudinal registration algorithm [7]. For each time point, within the predefined regions-of-interest including bilateral superior medial frontal areas and anterior cingulate cortex from the Automated Anatomical Labeling (AAL) template, group-level independent component analysis (gICA) is conducted to decompose multi-subject data in this area into eight spatially independent components based on the inherent spatiotemporal organizations. Of note, we vary the component number from 2 to 15, the 8-component setting got the best parcellation result. Then, we apply majority voting to generate non-overlapping subdivisions. Finally, we obtain the functional parcellation for each of the above seven time points. With visual inspection, only the components with the mirrored spatial distributions are merged into the same subregion.

We found that, even for the newborns, the functional subdivisions already exist in the mediofrontal areas that resemble the adult pattern (Fig. 2A). The changes along development in the subdivisions' pattern are minimal (Figs. 2B-G), indicating that the frontal subdivisions of the triple networks have already been formed and that the functional segregation is stable during the first two years of life. Of note, the parcellation was also done with adult data; all the results are quite similar to the adult result. Next, we examine the interconnectivity among the subdivisions to evaluate the functional integration.

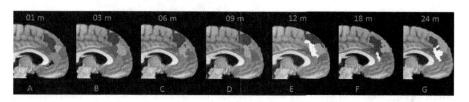

Fig. 2. Consistent parcellation of the mediofrontal area and its stable developmental pattern from 1 to 24 months (A-G). A dark blue area is part of the ECN, both red and purple areas belong to the SN, and a green area is within the DMN. Only a white area is emerging since 12 months old. (Color figure online)

We use the peak coordinates from the 24-month parcellation result as seeds for inter-subregion Pearson's correlation-based FC analysis for each subject at each time point. The individual FC matrices are then averaged across the same time point to produce a group-level FC matrix. To further examine the functional integration, based on all group-level FC matrices, modularity analysis is conducted to detect the strongly interconnected subregions as a module or community and separate those regions with weak FC into different modules. Modularity index is calculated to measure the development of community structure.

Similar to the stable spatial discreteness of the subdivisions in Fig. 2, these subdivisions have a quite stable modular structure (Fig. 3A). This confirms the hypothesis on the functional neuroanatomy of the triple networks in this area, that three dorsal medial subregions and three cingular subregions are separated but each with strong internal connections, and another subregion is singled out. As the infants grow up to 24 months old, this modular pattern becomes the same as the adult's (Fig. 3B).

Group-level modularity developmental trajectory indicates a non-monotonic, U-shape trend (Fig. 3C). Further investigation of the FC pattern reveals a dip at ~6 months old which is contributed by the overall FC increment at all links. After 9 months old, inter-modular FCs are decreased, but intra-modular FCs are preserved (Fig. 3D). Such overall early increased inter-modular FCs could be the consequence of myelination, dendritogenesis as well as prolonged synaptogenesis in such a higher cognitive-related area. These processes may be followed by a much dominant pruning process that helps to shape efficient brain connectivities.

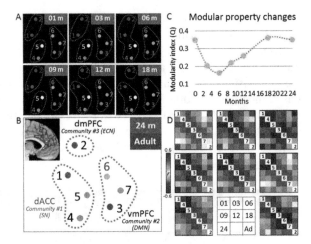

Fig. 3. Changes of modular structure within the first two years of life. Subplots A and B show the developing community structures among the mediofrontal area. The developmental trajectory of the group-level modularity follows U-shape with a turning point at ~6 months (C-D).

Besides local FC, based on group-level long-range FC analysis, we discover that different networks have distinct maturation speeds, with different timelines to reach their respective far-end brain regions (in an order of SN, DMN, and ECN, as shown in Fig. 4).

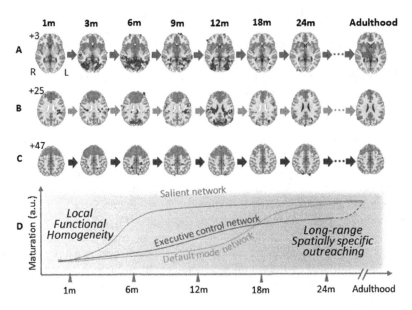

Fig. 4. Developing triple networks in the first two years of life defined by different far-end FC outreaching speeds. A: SN; B: DMN; C: ECN; and D: the proposed developing model.

2.2   Individual Prediction of Later Cognitive Performance

In addition to proposing a developing model for the triple networks at the group level, we further examine the feasibility of individualized prediction of the cognitive ability at the later age (such as 4 years old in this study).

The uniqueness of our data is that, it consists of *not only* neuroimaging data, *but also* various cognitive scores based on four scales evaluated after the infants grew up. These scales measure *(1)* visual reception ability, *(2)* fine motor ability, *(3)* receptive language ability, and *(4)* expressive language ability. They comprehensively measure a child's high-level visual, motor, language and memory abilities. Such wide spectrums of abilities are mediated by the high-level brain functional systems, particularly the triple networks. We further use a unified metric (i.e., early learning composite, ELC) fusing all the scores so that it generally indicates the overall cognitive performance. Individual-level triple-network FC and its developmental trajectory are assessed with respect to different ELC values.

There are totally 25 infants received the cognitive tests at 4 years old. There are no significant differences ($P > 0.05$) in gender, gestation age, and birth weight (Table 1) between the low and high ELC groups.

Table 1. Demographic information of all infants with the cognitive testing scores.

Variables	Low	High	Total
Total	13	12	25
Male	7	5	12
Female	6	7	13
ELC Mean ± STD	98.8 ± 13.3	129.9 ± 7.8	113.8 ± 19.2
Range	70–114	117–142	70–142
Gestation Age Mean ± STD (d)	278.8 ± 7.1	275.6 ± 7.8	277.3 ± 7.5
Birth Weight Mean ± STD (g)	3393.1 ± 366.6	3480.8 ± 519.5	3435.2 ± 439.2

ELC: early learning composite score; Low: the children with ELC scores lower than the median level; High: those with ELC scores higher than the median level.

To quantify *short-range* (within the mediofrontal area) and *long-range* FC (between the mediofrontal area and the other part of brain regions), we develop two novel metrics, namely, **modularity index (MI)** and **outreaching index (RI)**. MI is defined by the averaged intra-modular FC minus by the averaged inter-modular FC (Fig. 3B). RI is defined for each of the triple networks corresponding to the seed region #1 (for SN), #2 (for ECN), and #3 (for DMN) (Fig. 3B), calculated by Pearson's correlation of the whole brain gray matter FC pattern between each individual's FC map and the group-averaged adult FC map (derived by seed correlation based on 30 healthy college students [8] (aged 24 ± 2.41, 15 females)).

We first investigate *longitudinal developmental curves* of the MI and RI (see Fig. 5) for exemplary subjects with high or low ELC score. The curves for three infants (two with high cognitive performance and one with low performance) with complete seven scans are highlighted (for RI of the DMN only two infants were shown because the other one was detected as the outlier). For MI, we found similar trends with early decreased functional specialty at ~6 months, and the subject with lower cognitive performance has a delayed MI curve (Fig. 5A). For RI, SN has less individual variability than that for ECN and DMN (Fig. 5B), but they all have generally nonlinearly increased patterns. ECN's RI curves have larger individual variability (Fig. 5C). DMN's RI curves for two infants with high and low cognitive performance respectively show quite different trends (Fig. 5D). Collectively, these results demonstrate that individualized longitudinal analysis is important for understanding the highly heterogeneous cognitive ability in later time. The curve shape analysis with the newly proposed metrics (MI and RI) can reveal valuable predictive information.

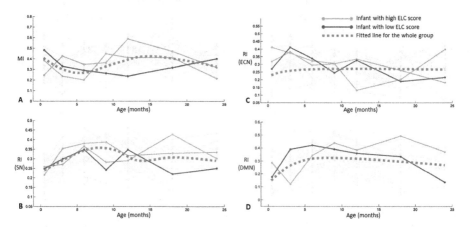

Fig. 5. Individual developmental trajectories of the MI and RI. RI curves are plotted for triple networks, separately. The outliers (outside 3 × STD) are removed. MI: modularity index; RI: outreaching index; ELC: early learning composite score.

We then investigate whether early brain connectomics can predict later behavior performance. We use the proposed highly informative within-mediofrontal FC *at birth* to *individually* predict the ELC score at 4-years old. We specifically use the neonatal brain connectomics to predict later high-level cognitive performance because it is more important to detect any abnormality as early as possible. Specifically, we extract inter-regional FC within the medial frontal region as features and use high and low ELC scores as labels, with a simple classification framework based on a linear support vector machine (SVM) and leave-one-out cross-validation, to conduct individual cognitive ability prediction. We further calculate the area under a receiver operating characteristic curve (AUC) to evaluate the predictability at each time point. We found that the early-stage prediction achieved high accuracy (with AUC of 0.72). This indicates that, although high-level cognitive-related brain region is less developed in neonates compared with other primary functions, the early high-level cognitive-related FC pattern within the medial frontal region could have predictive ability to indicate later cognitive performance. Both genetic and environmental factors during the period of gestation could contribute to such a finding.

3   Discussion and Conclusions

This is the first *in vivo* developmental study on healthy infants' three most important high-level brain functional systems. The functionally and clinically important triple network model has been, for the first time, extended from adulthood to early infancy. A novel developing triple network model, with the three high-level cognition-related networks having different maturation trajectories, is thus proposed. This is the first report on different high-level networks maturing with different trajectories. The SN may be related to attention to the salient stimulus; an infant may such abilities quite quickly. In addition, from the evolutional viewpoint, alertness and fast responses to the salient (mostly aversive) events could be one of the most necessary basic abilities for animals to survive. On the other hand, consciousness, self-recognition, executive control, attention maintenance and working memory abilities mediated by the DMN and ECN are believed to be a human privilege, which could have prolonged learning processes. From the neuroanatomical viewpoint, the DMN and ECN involve more long-range connections, while the SN is mostly located at insulo-frontal areas containing shorter connections. This could also be a reason for their different developmental trajectory. With gradual long-range myelination, long-range FCs will become more efficient.

We discovered a U-shape modular structure developmental curve which suggests that 6–9 months old is a pivotal period for the early development of high-level cognitive functions, during which numerous microscopic processes simultaneously occur. Interestingly, 6–9 months old is the same period as T1-weighted MRI contrast reversion occurs. During this period, within-mediofrontal FCs largely increase and result in a dense FC pattern; after that, the "redundant FC pattern" becomes more structured towards efficient information processing and functional specialization.

At last, we design a framework that utilizes both individual triple-network developmental trajectory and two quantitative metrics (MI and RI); and have conducted a preliminary individual cognitive performance prediction for neonates. The early FC patterns were found to be informative to predict later cognitive ability, which emphasizes the importance of early monitoring of the triple networks, especially their converging area, i.e., medial prefrontal region. Collectively, we suggest that the functional segmentation and integration in the medial prefrontal cortex might be largely determined at birth, while the relatively heterogeneous triple-network developmental trajectories may indicate a non-trivial environmental effect. This paper will broaden our knowledge on early neurodevelopment; most importantly, it demonstrates a promising future for individualized development monitoring for early intervention.

4   Future Works and Clinical Implications

The limited sample size due to the initial phase of baby connectome project can be alleviated when more infants are involved. Nevertheless, this work provides a feasible framework and indicates promising future for *early* individualized prediction of *later* cognitive ability based on baby connectomics. Although such prediction is currently based on a healthy cohort, in the future, the developing triple network model can be easily applied to screening high-risk infants or early detection of psychiatric disorders.

References

1. Gogtay, N., Giedd, J.N., Lusk, L., Hayashi, K.M., Greenstein, D., et al.: Dynamic mapping of human cortical development during childhood through early adulthood. Proc. Natl. Acad. Sci. USA **101**, 8174–8179 (2004)
2. Menon, V.: Large-scale brain networks and psychopathology: a unifying triple network model. Trends Cogn. Sci. **15**, 483–506 (2011)
3. Casey, B.J., Tottenham, N., Liston, C., Durston, S.: Imaging the developing brain: what have we learned about cognitive development? Trends Cogn. Sci. **9**, 104–110 (2005)
4. Gao, W., Alcauter, S., Elton, A., Hernandez-Castillo, C.R., Smith, J.K., et al.: Functional network development during the first year: relative sequence and socioeconomic correlations. Cereb. Cortex **25**, 2919–2928 (2015)
5. Casey, B.J., Pattwell, S.S., Glatt, C.E., Lee, F.S.: Treating the developing brain: implications from human imaging and mouse genetics. Annu. Rev. Med. **64**, 427–439 (2013)
6. Fjell, A.M., Walhovd, K.B., Brown, T.T., Kuperman, J.M., Chung, Y., et al.: Multimodal imaging of the self-regulating developing brain. Proc. Natl. Acad. Sci. USA **109**, 19620–19625 (2012)
7. Wu, G., Jia, H., Wang, Q., Shen, D.: SharpMean: groupwise registration guided by sharp mean image and tree-based registration. Neuroimage **56**, 1968–1981 (2011)
8. Chen, B., Xu, T., Zhou, C., Wang, L., Yang, N., Wang, Z., et al.: Individual variability and test-retest reliability revealed by ten repeated resting-state brain scans over one month. PLoS One **10**, e0144963 (2015)

Measuring Brain Connectivity via Shape Analysis of fMRI Time Courses and Spectra

David S. Lee, Amber M. Leaver, Katherine L. Narr, Roger P. Woods,
and Shantanu H. Joshi[✉]

Ahmanson-Lovelace Brain Mapping Center, Department of Neurology,
University of California, Los Angeles, CA, USA
s.joshi@g.ucla.edu

Abstract. We present a shape matching approach for functional magnetic resonance imaging (fMRI) time course and spectral alignment. We use ideas from differential geometry and functional data analysis to define a functional representation for fMRI signals. The space of fMRI functions is then equipped with a reparameterization invariant Riemannian metric that enables elastic alignment of both amplitude and phase of the fMRI time courses as well as their power spectral densities. Experimental results show significant increases in pairwise node to node correlations and coherences following alignment. We apply this method for finding group differences in connectivity between patients with major depression and healthy controls.

1 Introduction

Patterns of activation in the brain arising from task-based or resting state function magnetic resonance imaging (rfMRI) acquisitions are actively being investigated as potential biomarkers for pathology and healthy development of the brain. Often, network structures in the brain are defined using correlations in spontaneous low-frequency activity from BOLD fMRI signal across different brain areas, either targeting a given region's fMRI timecourse [1,14] or, less frequently, its power spectrum [12]. Implicit in the computation of correlations and coherence is the linear, one-to-one correspondence between the time series or the spectra, both across regions (i.e. for node to node correlations) and across subjects. This zero-lag assumption may not always hold true due to confounding effects by neuronal processes, synchronicity between different brain states, physiological noise, or even motion across subjects. Conversely, improved estimation of phase lags may also be informative when inferring directionality in network connections using fMRI data or in comparing the spectral content of fMRI timecourses across different brain regions. Recently researchers have proposed several ideas that compute the extremum of the cross-covariance [8] or perform a frequency-phase analysis [3] to discover this lag structure in rfMRI connectivity.

In our work, we adopt a functional data analysis approach to account for both the amplitude (peaks/valleys) changes and phase (time or frequency) delays

© Springer International Publishing AG 2017
G. Wu et al. (Eds.): CNI 2017, LNCS 10511, pp. 125–133, 2017.
DOI: 10.1007/978-3-319-67159-8_15

when inferring brain connectivity. Here, we would like to define functional representations of time courses or their spectra and use the functional shape information to align or match them across regions of interests or nodes, or across subjects. This *shape* alignment or matching is performed under a Riemannian metric that naturally gives rise to the connectivity measure that takes both the amplitude and phase into account. Specifically, we use the square root velocity function (SRVF) [4,5,10] to perform functional shape registration [11] of fMRI data. The novelty of our work includes two aspects; a new application of the functional data analysis framework to rfMRI signals, and for the first time we perform functional registration of rfMRI time courses and spectra using the nonlinear geometry of a function manifold. To our knowledge this has not been done before. To summarize, the contributions of this paper are as follows: (i) analysis of fMRI time courses and spectra using a functional data analysis framework, (ii) elastic shape matching of fMRI signals that enables the analysis of both amplitude and phase changes in fMRI across regions, and lastly (iii) the use of group level connectivity analysis for detecting changes in patterns of aligned fMRI signals across populations.

2   FMRI Shape Analysis of Time Courses and Spectra

In this section we describe the functional data analysis approach for analyzing fMRI signals and their spectra. Briefly, functional data analysis (FDA) has been widely applied to several problems in both computer vision and statistics [7,9,13]. The main idea is to define an object by a functional representation $f : I \to \mathbb{R}$, where I is the domain of the function. The function f is assumed to be square-integrable and thus is considered as an element of an infinite-dimensional Hilbert space. This Hilbert space naturally allows the $\mathbb{L}^2$ inner product $\langle f_1, f_2 \rangle = \int_I f_1(t) f_2(t) dt$ that also serves as a metric for finding distances between functions. This framework can then be used to perform statistical modeling including regression, prediction and classification.

2.1   Elastic Functional Data Analysis of fMRI Signals Using SRVFs

fMRI signal representation: Here, we describe the functional representation for fMRI signals in brief. For more details the reader is referred to [11]. For a given fMRI time course signal $f : I \equiv [0, 1] \to \mathbb{R}$, and its velocity $\dot{f}(t) = \frac{df}{dt}$ and magnitude $|\dot{f}(t)|$, we define its functional representation by the square-root velocity field (SRVF) map q given by,

$$q : [0, 1] \to \mathbb{R}, \ q(t) = \frac{\dot{f}(t)}{\sqrt{|\dot{f}(t)|}}. \tag{1}$$

For an absolutely continuous f, the SRVF transformation ensures that q is square integrable. The set of SRVFs is then given by $\mathbb{L}^2([0, 1], \mathbb{R})$, which is a Hilbert space. The original fMRI signal can be recovered by $f(t) = f(0) + \int_0^t q(\tau)|q(\tau)|d\tau$.

The SRVF mapping is invertible up to a given $f(0)$. We assume $f(0) = 0$ as the initial condition of the fMRI signal at time $t = 0$. It is noted that the domain I is defined as the interval $[0, 1]$ for all signals in the population. We will use the same notation for denoting the frequency domain spectrum of the fMRI signal given by its power spectral density (PSD) estimate. In this case, the notation for time domain t is changed to that of the frequency domain ν. Then using q for the SRVF mapping, we have $q : [0, 1] \to \mathbb{R}$, $q(\nu) = \frac{\dot{p}(\nu)}{\sqrt{|\dot{p}(\nu)|}}$, where p is the PSD function of f. With a slight abuse of notation, we will denote q for the SRVF mapping of both the fMRI time course and the PSD.

fMRI temporal domain and spectral domain reparameterization: To account for temporal shifts and spectral phase lags, we now define the notion of time and frequency reparameterization. This idea is closely related to the parameterization (speed) of the underlying domain on which the function f or ν is defined. For example, increasing the speed of the parameterization results in local shrinking of the domain, whereas reducing the speed of the parameterization results in local expansion of the parameterization domain. This behavior can be modeled by a warping function $\gamma : I \to I$, where $\dot{\gamma} > 0, \forall t \in I$; γ being a diffeomorphism. Thus to change the temporal parameterization, one can simply compose f with γ as $f \circ \gamma$. In the SRVF domain, this is given by

$$q \cdot \gamma = \frac{(\dot{f} \circ \gamma)\dot{\gamma}}{\sqrt{|(\dot{f} \circ \gamma)\dot{\gamma}|}} = \frac{(\dot{f} \circ \gamma)}{\sqrt{|(\dot{f} \circ \gamma)|}}\sqrt{\dot{\gamma}} = (q \circ \gamma)\sqrt{\dot{\gamma}}. \tag{2}$$

We denote the set of all possible γ functions as Γ and emphasize that incorporating domain warping via γ functions enables elastic shape matching of fMRI functions.

Elastic Riemannian metric for SRVFs of fMRI signals: To compare functions and compute distances between them, we define the notion of a metric on the space of q functions. Before analysis of fMRI signals, one usually standardizes the signal by obtaining a z score of f as given by $\tilde{f} = \frac{f - \bar{f}}{\sigma}$, where $\bar{f}$ is the mean value of f and σ is the standard deviation. One can impose an analogous unit length constraint on the q function by obtaining $\tilde{q} = \frac{q}{||q||}$. This unit length transformation forces q to lie on a Hilbert sphere denoted by $\mathcal{Q}$. Formally, the space $\mathcal{Q}$ is defined as $\mathcal{Q} \equiv \left\{ q \in \mathbb{L}^2 | \int_0^1 (q(s), q(s))_{\mathbb{R}^2} ds = 1, q(s) : [0, 1] \to \mathbb{R}^2 \right\}$. Then one can define the Riemannian metric on the tangent space of this sphere $T_q(\mathcal{Q})$. An important feature of the SRVF representation for fMRI signals is that an elastic Riemannian metric, which is invariant to the domain reparameterization, is reduced to the $\mathbb{L}^2$ metric [11] and given by $d(f_1, f_2) = ||q_1 - q_2||$. Therefore, for any two SRVFs given by $q_1, q_2 \in \mathbb{L}^2$ and $\gamma \in \Gamma$, we have $||q_1 \cdot \gamma - q_2 \cdot \gamma|| = ||q_1 - q_2||$. This property allows us to solve the problem of registration in an efficient, invariant manner.

2.2   fMRI Alignment and Registration:

Next, we enable comparisons between functions via elastic geodesics between them. Since the space $\mathcal{Q}$ is a Hilbert sphere, the geodesic between two points (shapes) q_1 and q_2 can be expressed analytically as,

$$\chi_t(q_1; v) = \cos\left(t\cos^{-1}\langle q_1, q_2\rangle\right) q_1 + \sin\left(t\cos^{-1}\langle q_1, q_2\rangle\right) v, \tag{3}$$

where $t \in [0, 1]$ and the initial tangent vector $v \in T_{q_1}(\mathcal{Q})$ is given by $v = q_2 - \langle q_1, q_2\rangle q_1$. Then the geodesic distance between the two shapes q_1 and q_2 in $\mathcal{Q}$ is given by

$$d(q_1, q_2) = \int_0^1 \sqrt{\langle \dot{\chi}_t, \dot{\chi}_t\rangle} dt. \tag{4}$$

To find the elastic geodesic distance, we simply minimize Eq. 4 as $d_{\text{elastic}}(q_1, q_2) = \min_{\gamma \in \Gamma} d(q_1, q_2 \cdot \gamma)$. In estimating the elastic geodesic, the optimal reparameterization $\hat{\gamma}$ can be efficiently found as the minimizer $\hat{\gamma} = \text{argmin}_\gamma \left(\int_0^1 ||q_1 - \gamma \cdot q_2||^2 dt\right)$. In practice, we use dynamic programming to find the optimal $\hat{\gamma}$. The phase difference between two functions is encoded by the warping function $\hat{\gamma}$ resulting from the alignment.

Group analysis and statistics of fMRI signals: For statistical analysis of fMRI signals and spectra, we introduce the notion of the Karcher mean [6]. Given a collection of functions $f_1, f_2, \cdots, f_n$, let $q_1, q_2, \cdots, q_n$ denote their SRVFs, respectively. The Karcher mean is then computed by an iterative procedure: initialize the mean function μ^k at an iteration k and solve for

$$\hat{\gamma}_i^{k+1} = \arg\inf_{\gamma \in \Gamma} \|\mu^k - (q_i \circ \gamma)\sqrt{\dot{\gamma}}\|, i = 1, 2, \cdots, n, \tag{5}$$

$$\mu^k = \frac{1}{n}\sum_{i=1}^n (q_i \circ \hat{\gamma}_i^{k-1})\sqrt{\dot{\hat{\gamma}}_i^{k-1}}. \tag{6}$$

One can use this mean function as a template for aligning all the functions in the group. This enables one to compare fMRI signals across population.

3   Results

In this section we describe experimental results that show improvement in pairwise node-to-node correlation and coherence as well as group differences in connectivity between healthy controls and patients with major depressive disorder (MDD). 70 patients (34M/36F, mean age 43 years) with MDD and 36 healthy volunteers (17M/19F, mean age 39 years) underwent fMRI imaging on a 3T Siemens Allegra scanner (TR = 2 s, TE = 30 ms, flip angle = 70, $3.4\times3.4\times5$ mm^3 resolution). We used FSL [4] to perform slice-timing correction, motion correction, and high pass filtering. The fMRI scans were then filtered using ICA based denoising and registered to the T1-weighted anatomical MPRAGE scan. All images were normalized to the MNI standard space using SPM [1]. We parcellated the fMRI images using the Craddock functional atlas [2] and focused our analysis on 18 seed regions chosen based on their relevance to depression. They included the subgenual, rostral, and dorsal anterior cingulate cortex (ACC), bilateral amygdala and overlapping anterior hippocampus (am hp), bilateral dorsolateral prefrontal cortex (DLPFC), bilateral thalamus (Th), posterior cingulate cortex (PCin), and bilateral precuneus (PreCun). Additional regions less relevant to depression were chosen as control nodes; these included bilateral primary visual cortex along the calcarine sulcus (Visual 1 & 2).

3.1   Visualization of Elastic Functional Alignment

Figure 1 shows examples of time course and spectral alignment of fMRI signals across regions and within-subject. In Fig. 1, left, non-elastic (Panels A, C, and E $\gamma = $ identity $= t$) and elastic matching (Panels B, D, and F) of two time courses and PSDs are compared. In non-elastic matching (Panel A), the two curves are analyzed at each time or frequency, as represented by the vertical black lines. In contrast, in elastic matching (Panel B), similar features of the two curves are aligned. As a result, the peaks and valleys of the two time series in panel F are aligned after elastic matching.

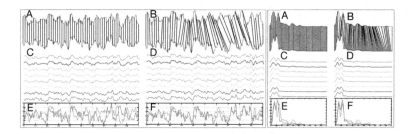

Fig. 1. Within-subject registration of resting state time courses (Left), and power spectral densities (PSDs, Right). A: non-elastic matching between the top and bottom fMRI signals. B: elastic matching shown by corresponding lines. C: non-elastic and D: elastic geodesics between the two time signals (Left), and PSDs (Right). E: overlay of top and bottom signals before matching. F: overlay after matching.

Alignment across subjects allows estimation of a template, which can serve as a reference for group analyses. Figure 2 shows the mean PSDs of the population ($N = 106$) in 18 regions of interest without and with alignment. While the non-elastic mean seems to capture a single low-frequency feature, the elastic mean identifies features across the frequency range. Finally, Fig. 3 shows PSDs of the dorsal anterior cingulate and dorsolateral prefrontal cortex aligned to the average PSD for those ROIs for $N = 50$ randomly selected subjects. Elastic alignment yields spectra with distinct peaks at both low and high frequencies, mirroring the mean spectra displayed in Fig. 3.

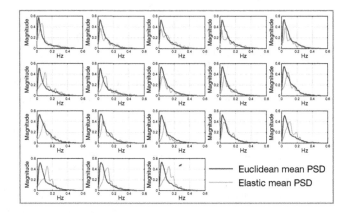

Fig. 2. Mean PSDs for 18 regions of interest.

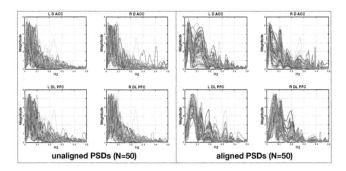

Fig. 3. Left panel: Unaligned PSDs overlaid; each line corresponds to data from a single subject. Right panel: PSDs after elastic alignment to the mean shape. ROIs shown are the left and right dorsal anterior cingulate and the left and right prefrontal cortex.

3.2 Measuring Brain Connectivity After Elastic fMRI Alignment

Pairwise node to node connectivity measures were obtained by computing the Pearson correlation between time series, coherence between power spectra, and elastic geodesic distances between time series. We remind the reader that the elastic geodesic distance measures the difference between the fMRI signals, whereas measures such as correlation and coherence measure the closeness or agreement between the fMRI signals.

Increases in Measures of Correlation and Coherence. As expected we observed increases in correlation and coherence after elastic functional alignment. The effect of alignment was evaluated by comparing correlations of time series and coherences of the PSDs for each pair of nodes across all subjects with and without alignment. Figure 4 shows a signed value of Cohen's d computed as $d(x, x_{aligned}) = \frac{\mu_x - \mu_{x_{aligned}}}{\sigma_{x,x_{aligned}}}$ at each node, where x and $x_{aligned}$ represent the correlations or coherence before and after alignment. The measures were found to be consistently higher in majority of connections following alignment. Connectivity among visual cortex was high prior to alignment; therefore changes in these connections were modest.

Population Analysis of Connectivity Patterns. Lastly, group differences in functional connectivity between patients and controls were examined using a linear model covaried with age and gender. In addition to the widely used correlation and coherence, we also used the geodesic distances between nodes and the deviation of the warping function γ from the *identity* as potential measures of connectivity. Figure 5 shows connectivity differences between patients with MDD and healthy controls.

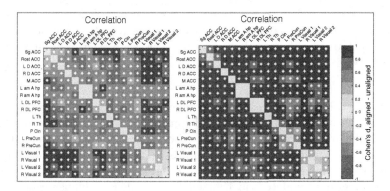

Fig. 4. Signed Cohen's d shown at each node, with a white asterisk denoting significance at $p < 0.05$ (uncorrected). ACC-Anterior Cingulate Cortex; Am A hp-Amygdala/Anterior hippocampus; Cin-Cingulate; D-Dorsal; L-Left; P-Posterior; PreCun-Precuneus; R-Right; Rost-Rostral; Sg-Subgenual.

Several features in group analysis were observed with alignment. Some connections represented by correlation or coherence were maintained with alignment, for example, between anterior cingulate and dorsal prefrontal cortex. Importantly, effect sizes of the group differences increased for existing connections after alignment. Not all connections were preserved after alignment. For example, the correlations between the precuneus and thalamus were weakened after alignment. On the other hand, the correlation and coherence between posterior cingulate and thalamus were increased after alignment.

4   Discussion

We proposed an elastic shape matching approach for the analysis of fMRI time series and PSDs. It is worth noting that several significant inter-node connections shown by elastic geodesic and gamma distances coincided with those shown by

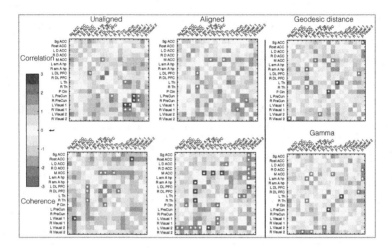

Fig. 5. Group differences between patients with MDD and healthy controls. Colormap shows the t-statistic with controls > patients. The white asterisk denotes pairwise significance at $p < 0.05$ (uncorrected).

correlation or coherence. For example, significantly higher correlation after alignment, and coherence before and after alignment between thalamus and medial anterior cingulate are captured by lower geodesic and gamma distances. On the other hand, the lower correlation in controls between precuneus and thalamus is encoded by a higher gamma distance, whereas lower coherence in controls between posterior cingulate and thalamus is encoded by higher geodesic distances. This suggests that the geodesic and gamma distances from elastic alignment may serve as additional representations of connectivity. While we observed an increase in correlation and coherence between fMRI signals following alignment, further validation will be necessary to explore the clinical utility of this approach. In addition, we anticipate enhanced effect of alignment in task-based fMRI where neurobiological signals are coherent with tasks.

Acknowledgments. This research was supported by the NIH/NIAAA award K25AA024192, and the NIH/NIMH awards K24MH102743 and U01MH110008.

References

1. Baria, A.T., Baliki, M.N., Parrish, T., Apkarian, A.V.: Anatomical and functional assemblies of brain bold oscillations. J. Neurosci. **31**(21), 7910–7919 (2011)
2. Craddock, R.C., James, G.A., Holtzheimer, P.E., Hu, X.P., Mayberg, H.S.: A whole brain fMRI atlas generated via spatially constrained spectral clustering. Hum. Brain Mapp. **33**(8), 1914–1928 (2012)
3. Goelman, G., Dan, R., Růžička, F., Bezdicek, O., Růžička, E., Roth, J., Vymazal, J., Jech, R.: Frequency-phase analysis of resting-state functional MRI. Scientific Reports **7** 2017

4. Joshi, S.H., Klassen, E., Srivastava, A., Jermyn, I.: A novel representation for Riemannian analysis of elastic curves in R^n. In: IEEE Conference on Computer Vision and Pattern Recognition (CVPR), pp. 1–7. IEEE (2007)
5. Joshi, S.H., Klassen, E., Srivastava, A., Jermyn, I.: Removing shape-preserving transformations in square-root elastic (SRE) framework for shape analysis of curves. In: Yuille, A.L., Zhu, S.-C., Cremers, D., Wang, Y. (eds.) EMMCVPR 2007. LNCS, vol. 4679, pp. 387–398. Springer, Heidelberg (2007). doi:10.1007/978-3-540-74198-5_30
6. Karcher, H.: Riemannian center of mass and mollifier smoothing. Commun. Pure Appl. Math. **30**, 509–541 (1977)
7. Kneip, A., Ramsay, J.O.: Combining registration and fitting for functional models. J. Am. Stat. Assoc. **103**(483), 1155–1165 (2008)
8. Mitra, A., Snyder, A.Z., Hacker, C.D., Raichle, M.E.: Lag structure in resting-state fMRI. J. Neurophysiol. **111**(11), 2374–2391 (2014)
9. Ramsay, J.O., Li, X.: Curve registration. J. Roy. Stat. Soc. Ser. B (Statistical Methodology) **60**(2), 351–363 (1998)
10. Srivastava, A., Klassen, E., Joshi, S.H., Jermyn, I.H.: Shape analysis of elastic curves in Euclidean spaces. IEEE Trans. Patt. Anal. Mach. Intell. **33**, 1415–1428 (2011)
11. Srivastava, A., Wu, W., Kurtek, S., Klassen, E., Marron, J.S.: Registration of functional data using Fisher-Rao metric. arXiv:1103.3817v2 (2011)
12. Sun, F.T., Miller, L.M., D'Esposito, M.: Measuring interregional functional connectivity using coherence and partial coherence analyses of fMRI data. Neuroimage **21**(2), 647–658 (2004)
13. Tang, R., Muller, H.-G.: Pairwise curve synchronization for functional data. Biometrika **95**(4), 875–889 (2008)
14. Zuo, X.-N., Di Martino, A., Kelly, C., Shehzad, Z.E., Gee, D.G., Klein, D.F., Castellanos, F.X., Biswal, B.B., Milham, M.P.: The oscillating brain: complex and reliable. Neuroimage **49**(2), 1432–1445 (2010)

Topological Network Analysis of Electroencephalographic Power Maps

Yuan Wang$^{(\boxtimes)}$, Moo K. Chung, Daniela Dentico, Antoine Lutz,
and Richard J. Davidson

University of Wisconsin, Madison, USA
yuanw@stat.wisc.edu

Abstract. Meditation practice is a non-pharmacological intervention that provides both physical and mental benefits. It has generated much neuroscientific interest in its effects on brain activity. Spontaneous brain activity can be measured by electroencephalography (EEG). Spectral powers of EEG signals are routinely mapped on a topographic layout of channels to visualize spatial variations within a certain frequency range. In this paper, we propose a *node-based network filtration* to model the spatial distribution of an EEG topographic power map via its dynamic local connectivity with respect to a changing scale. We compare topological features of the network filtrations between long-term meditators and mediation-naïve practitioners to investigate if long-term meditation practice changes power patterns in the brain.

1 Introduction

Meditation is a set of mental training regimes widely practiced for its claimed benefits to physical and mental health. The investigation of spontaneous brain activity during resting state or practice, is a sensitive approach to identify neuroplastic changes induced by meditation practice [2]. Electroencephalogram (EEG) is an important imaging modality for exploring the neuroplastic effects of meditation under various experimental conditions. In these studies, spectral powers of EEG signals are routinely mapped on a topographic layout of channels to visualize spatial variations within a certain frequency range. Topographic difference in spectral powers indicates configuration change in the brain's active neuronal sources. It is thus important to establish a statistical framework for comparing topographic power maps in the study of neuroplastic effect of long-term meditation practice.

Statistical inference of EEG topographic power maps is typically based on the mass univariate approach with multiple node-level testing [7]. This approach does not account for the network topology in the topography of the power map. Alternative statistical methods more commonly applied to electric potential maps include microstate analysis [1] and cluster-based inference [8]. But microstate analysis uses a global dissimilarity index based on node-level mean

© Springer International Publishing AG 2017
G. Wu et al. (Eds.): CNI 2017, LNCS 10511, pp. 134–142, 2017.
DOI: 10.1007/978-3-319-67159-8_16

difference and variance rather than network topology in the topography. Cluster-based methods often require threshold selection which may result in bias and inconsistency [5,9].

In this paper, we propose a *node-based network filtration* for modeling the spatial distribution of an EEG topographic power map. Each EEG power map is modeled as an undirected network on a triangulation of the map, with node weights defined from denoised frequency powers. We binarize the network by thresholding the node weights, and obtain the network filtration - a nested sequence of binary networks - as we vary the threshold. A topological feature of the filtration is then incorporated in a permutation test for group difference between the maps. Simulation studies show evidence that the proposed framework is robust to scaling and translation of maps and sensitive to translation in opposite directions resulting in map spatial difference. The proposed framework is also applied to compare the topographic power maps of long-term meditators and meditation naïve practitioners.

The methodological contributions of this paper are: (1) we propose a node-based network filtration for quantifying the spatial distribution of an EEG topographic power map; (2) we use the node-based network filtration to make spatial comparison of two groups of EEG power maps.

2   Methods

Our goal is to compare spatial distribution of EEG power maps in meditators and novices. We first briefly describe a spatial denoising procedure on a power map. We then characterize the spatial distribution of the denoised power map through a sequence of binary networks constructed on the map.

EEG topographic power map. Signal at each of the c observed EEG channels $v_1, v_2, \ldots, v_c$ is decomposed into frequency components by Fourier transform. The strengths of the frequency components within a certain range are measured by integrating the power spectral density (PSD). Here we estimate the PSD of the EEG signal at each channel by Welch's method of modified periodogram: divide a signal into overlapping segments and then average the modified periodograms computed on all the segments to obtain a PSD estimate with reduced variance than the usual periodogram [10]. We denote the topographic map of the PSDs at c EEG channels by $\boldsymbol{f} = (f_1, \ldots, f_c)$, where the index follows the EEG channel labels.

Spatial denoising. We then spatially denoise the topographic power map $\boldsymbol{f}$ of each subject at a particular frequency band. We model the topography of $\boldsymbol{f}$ as an undirected graph $\mathcal{G} = \{\mathcal{V}, \mathcal{E}\}$ with the node set

$$\mathcal{V} = \{v_i : i = 1, \ldots, c\}$$

of the c EEG channels and the edge set with no orientation

$$\mathcal{E} = \{(v_i, v_j) : v_i, v_j \in \mathcal{T_V}, v_i \sim v_j, i, j = 1, \ldots, c\},$$

where $\mathcal{T}_\mathcal{V}$ is the Delaunay triangulation built on $\mathcal{V}$ and $\sim$ denotes neighbors in $\mathcal{T}_\mathcal{V}$. Defining the graph Laplacian $L = (l_{ij})$ on $\mathcal{G}$ by

$$l_{ij} = \begin{cases} -a_{ij}, & v_i \neq v_j \text{ and } v_i \sim v_j \\ \sum_{k \neq i} a_{ik}, & v_i = v_j \\ 0, & \text{otherwise} \end{cases}$$

with the adjacency matrix $A = (a_{ij})$, there are up to c unique eigenvectors $\psi_1, \psi_2, \cdots, \psi_c$ satisfying

$$L\psi_j = \gamma_j \psi_j \qquad (1)$$

with $0 \leq \gamma_1 \leq \gamma_2 \leq \cdots \leq \gamma_c$. These eigenvectors are orthonormal, i.e., $\psi_i' \psi_j = \delta_{ij}$ - the Kronecker's delta. The first eigenvector is trivial: $\psi_1 = 1/\sqrt{c}(1, \ldots, 1)'$. All other eigenvalues and eigenvectors are analytically unknown and need to be numerically computed.

Once we obtain eigenvectors ψ_j satisfying (1) on the Delaunay triangulation $\mathcal{T}_\mathcal{V}$, the heat kernel estimate for the power map $\boldsymbol{f}$ is given by

$$\widehat{\boldsymbol{f}} = (\widehat{f}_1, \ldots, \widehat{f}_c) = K_\sigma * \boldsymbol{f} = \sum_{j=1}^{c} e^{-\gamma_j \sigma} \zeta_j \psi_j, \qquad (2)$$

where $K_\sigma = \sum_{j=1}^{c} e^{-\gamma_j \sigma} \psi_j \psi_j'$ is the discrete heat kernel and $\zeta_j = \boldsymbol{f}' \psi_j = \psi_j' \boldsymbol{f}, j = 1, \ldots, c$, are the Fourier coefficients with respect to the basis $\{\psi_1, \ldots, \psi_c\}$. The parameter σ is the heat kernel bandwidth and it modulates the extent of denoising.

Quantifying the spatial distribution of a power map. We define a node-weighted network on the map through $\mathcal{G} = \{\mathcal{V}, \mathcal{E}\}$, with the node weights

$$w_i = \widehat{f}_i, i = 1, \ldots, c,$$

assumed to be unique. With respect to an arbitrary threshold $\lambda \in \mathbb{R}$, we define a binary network

$$\mathcal{G}_\lambda = \{\mathcal{V}_\lambda, \mathcal{E}_\lambda\}$$

on $\mathcal{G}$, where

$$\mathcal{V}_\lambda = \{v_i \in \mathcal{V} : w_i \leq \lambda\}$$

and

$$\mathcal{E}_\lambda = \{(v_i, v_j) \in \mathcal{E} : \max(w_i, w_j) \leq \lambda\}.$$

Now let

$$\lambda_1 = w_{(1)} < \lambda_2 = w_{(2)} < \cdots < \lambda_c = w_{(c)}$$

be the order statistics of the unique node weights $w_1, w_2, \ldots, w_c$ of $\mathcal{G}$. Setting λ in the order of $\lambda_1, \lambda_2, \cdots, \lambda_c$ yields a sequence of subsets of $\mathcal{G}$:

$$\mathcal{G}_{\lambda_1} \subset \mathcal{G}_{\lambda_2} \subset \cdots \subset \mathcal{G}_{\lambda_c}, \qquad (3)$$

which we call a *node-based network filtration*.

Note that the filtration (3) is not affected by relabeling of the EEG channels, since the order statistics $\lambda_i = w_{(i)}, i = 1, \ldots, c$, remain the same regardless of the channel labels. Each $\mathcal{G}_\lambda$ in (3) consists of clusters of nodes; as λ increases, clusters appear and later merge with existing clusters. The pattern of changing clusters in (3) has the following key properties.

(1) For all $\lambda_i < \lambda < \lambda_{i+1}$, $\mathcal{G}_\lambda = \mathcal{G}_{\lambda_i}, i = 1, \ldots, c - 1$; in other words, the filtration (3) is maximal in the sense that no more $\mathcal{G}_\lambda$ can be added to it.

(2) As λ increases from λ_i to λ_{i+1}, only the node v'_{i+1} that corresponds to the weight λ_{i+1} is added in $\mathcal{V}_{\lambda_{i+1}}$.

(3) Define a local minimum (maximum) λ_i as

$$\lambda_i < \lambda_j \ (\lambda_i > \lambda_j), \forall \ v'_j \sim v'_i,$$

where v'_i and v'_j are nodes that correspond to the weights λ_i and λ_j. New cluster of nodes emerge in $\mathcal{G}_{\lambda_i}$ at a local minimum λ_i and merge with other clusters at a local maximum λ_i. Here we assume that we do not encounter the case where

$$\lambda_i < \lambda_j \text{ some } v'_j \sim v'_i \text{ and } \lambda_i > \lambda_j \text{ the other } v'_j \sim v'_i.$$

Properties (1) and (2) hold because the $\lambda_i, i = 1, \ldots, c$ account for all the unique node weights $w_i, i = 1, \ldots, c$. Property (3) holds for local minimum λ_i because all the neighboring nodes v'_j of v'_i are not included in $\mathcal{G}_{\lambda_i}$, hence v'_i emerges as a standalone cluster in $\mathcal{G}_{\lambda_i}$; for local maximum λ_i, clusters to which the v'_i are connected are joined by v'_i in $\mathcal{G}_{\lambda_i}$.

We illustrate the filtration (3) on a 6-channel EEG layout in the international 10–20 system (Fig. 1). We first build up the Delaunay triangulation over the 6-channel layout (Fig. 1). Node weights are the powers at the EEG channels. At each filtration value λ, we include the nodes and edges with weights less than or equal to λ. The clusters change as λ increases.

Topological permutation test. We use a topological feature to summarize the changing connectivity in the sequence of binary networks. The 0th Betti number β_0 counts the number of clusters in a network [4]. In this paper we define the 0th Betti function at $\lambda_1 < \cdots < \lambda_m$ as the sequence of 0th Betti numbers $(\beta_0^1, \ldots, \beta_0^m)$. For instance, the 0th Betti function in Fig. 1 corresponding to $\lambda = -1, 0, 0.5, 1, 2, 3$ is $(1, 1, 2, 1, 1, 1)$.

Same spatial distribution implies the same node-based network filtration, hence the same 0th Betti function. To statistically compare the spatial distribution of two groups of denoised power maps, we test the null hypothesis that there is no difference between the respective mean 0th Betti functions $\bar{\beta}_0^1$ and $\bar{\beta}_0^2$ of the node-based network filtrations of maps in Group 1 and 2:

$$H_0 : \bar{\beta}_0^1(\lambda) = \bar{\beta}_0^2(\lambda), H_1 : \bar{\beta}_0^1(\lambda) \neq \bar{\beta}_0^2(\lambda), \tag{4}$$

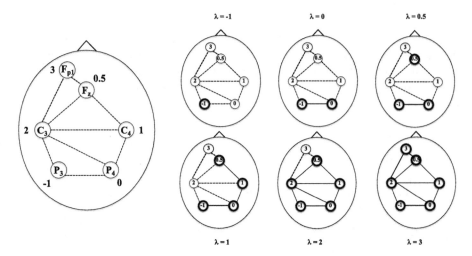

Fig. 1. Schematic of the filtration (3) on 6 weighted EEG channels in the international 10–20 system. (a) Large figure on left: The 6-channel layout with the corresponding Delaunay triangulation indicated by dashed lines. Node weights are the powers at the EEG channels. (b) Small figures on right: At each filtration value λ, we include the nodes and edges with weights less than or equal to λ. As the λ increases, more nodes and edges join in the filtration.

at fixed m filtration values $\lambda_1, \ldots, \lambda_m$. To test the null hypothesis (4), we first compute the ℓ_2 distance

$$\ell_2(\bar{\beta}_0^1, \bar{\beta}_0^2) = \sqrt{\sum_{i=1}^{m} (\bar{\beta}_0^1(\lambda_i) - \bar{\beta}_0^2(\lambda_i))^2}, \tag{5}$$

between the respective group means

$$\bar{\beta}_0^1 = (\bar{\beta}_0^1(\lambda_1), \ldots, \bar{\beta}_0^1(\lambda_m)) \text{ and } \bar{\beta}_0^2 = (\bar{\beta}_0^2(\lambda_1), \ldots, \bar{\beta}_0^2(\lambda_m))$$

of the 0th Betti functions of the node-based network filtrations characterizing the denoised power maps in Group 1 and 2. Then the labels of the two groups undergo repeated random exchanges. At each label exchange, the $\ell_2(\bar{\beta}_0^{1'}, \bar{\beta}_0^{2'})$ distance between the respective mean Betti functions $\bar{\beta}_0^{1'}$ and $\bar{\beta}_0^{2'}$ of the relabeled power maps. We take the proportion of the distances $\ell_2(\bar{\beta}_0^{1'}, \bar{\beta}_0^{2'})$ exceeding that of the observed distance $\ell_2(\bar{\beta}_0^1, \bar{\beta}_0^2)$ is taken as the p-value for the permutation test.

3   Simulations

We use simulations to evaluate how well the proposed topological permutation test detects difference in the spatial distribution of two groups of power maps.

A scaled or translated map has identical filtration as the original map after normalization. So the proposed test should stay robust under map scaling and translation with moderate noisy perturbations. It should also be sensitive to spatial difference between maps caused by translation in opposite directions.

We simulate two groups of noisy power maps by first defining the underlying function $z = (z_1, \ldots, z_{100})$ by

$$z_i = 3(1 - x_i)^2 e^{-(x_i^2 + y_i^2)} + 3e^{-((x_i-2)^2 + y_i^2)}, i = 1, \ldots, 100, \qquad (6)$$

with the Cartesian coordinates $(x_1, y_1), \ldots, (x_{100}, y_{100})$ sampled uniformly from the four quadrants of the $[-3,3] \times [-3,3]$ grid. We then define a transformation $z' = (z'_1, \ldots, z'_{100})$ of z through one of the following functions:

1. (scaling)
$$z'_i = 5z_i;$$

2. (translation)
$$z'_i = (z_i + 5);$$

3. (translation in opposite directions)

$$z'_i = (z_i \pm 5)$$

($+$ for $1 \leq i \leq 50$ and $-$ for $51 \leq i \leq 100$), which translates two halves of the map in opposite directions.

We add independent Gaussian noises $N(0, 0.1^2)$ to z and z' at the $(x_i, y_i), i = 1, \ldots, 100$, to create two groups of power maps $\{z_1, \ldots, z_5 : z_j = (z_{j1}, \ldots, z_{j100})\}$ and $\{z'_1, \ldots, z'_5 : z'_j = (z'_{j1}, \ldots, z'_{j100})\}$.

Under each transformation setting, this simulation procedure is repeated 500 times; for each simulation, the null hypothesis (4) is tested on the 2 groups of 5 samples through the proposed permutation test with 252 exact permutations. We reject the null when a p-value falls below 0.05. The rejection rates are 5%, 3% and 98% in each setting. The results provide numerical evidence that the proposed procedure for testing the difference between topographic maps stays robust under some scaling and translation and meanwhile is sensitive to translation in opposite directions. In comparison, the maximum t-statistic test has rejection rates of 9%, 6% and 99% in each setting. It is more sensitive than the proposed topological inference procedure in picking up non-topological difference between power maps.

4   Real Data Application

Data description. The aim of this application is to compare topological difference between frequency variations in the EEG signals of 24 meditation-naïve participatns (MNPs) and 24 long-term meditators (LTMs) of Buddhist meditation practices (approximately 8700 mean hours of life practice) during whole-night non-rapid eye movement (NREM) sleep divided into 3 cycles. The EEG

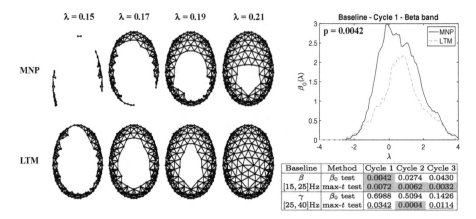

Fig. 2. Left: Filtrations of mean normalized power maps in the beta band in sleep cycle 1 under the baseline condition. Right top: Group mean β_0 functions with the p-value from the β_0 permutation test. Right bottom: The p-values of β_0 and maximum t-statistic permutation tests comparing MNPs and LTMs in the baseline session. The p-values below the Bonferonni threshold $0.05/6 = 0.0083$ corrected over 2 (frequency bands) $\times$ 3 (sleep cycles) = 6 tests for each method are shaded in gray.

signals were recorded with a 256-channel hdEEG system (Electrical Geodesics Inc., Eugene, OR). Signals bandpass filtered (1–50 Hz), and independent component analysis was used to remove ocular and muscle artifacts in the signals. More pre-processing details can be found in [3]. The participants undergo 3 sessions of recording: a baseline session, and one session each after two days of Vipassana (mindfulness) and Metta (compassion) meditations. We analyze the baseline session for unconfounded effect of long-term meditation practice. Also, we focus on the high-frequency bands β and γ of the EEGs since high frequency has been shown to positively correlate with meditation experience [6].

Topological permutation test. After heat kernel denoising with a moderate bandwidth $\sigma = 0.5$ for the noise level in the data, we normalize each power map by a z-score transformation across all channels. We then compare the normalized denoised power maps of the LTMs and MNPs in the high-frequency β (15–25 Hz) and γ (25–40 Hz) bands by the proposed permutation test. For β band in sleep cycle 1, the node-based network filtrations of the average normalized maps in both groups are shown in Fig. 2 (left). The closure of clusters is distinctly faster in the average LTM map as λ increases. Figure 2 (right top) shows the average β_0 functions of LTMs and MNPs in the β band of sleep cycle 1. The LTM function is below the MNP function throughout the range of λ values, meaning that on average the LTMs have fewer clusters than the MNPs. This implies that the LTM power maps having more coherent spatial distribution, as nodes with similar powers get connected in a smaller window of λ than those with more varied powers.

Comparison with maximum t-statistic test. The table of p-values in Fig. 2 provides comparison between results of the proposed and maximum t-statistic permutation test. The only place where the proposed test shows significant topological difference is the β band in sleep cycle 1, whereas the maximum t-statistic test shows significant difference between LTM and MNP in four out of six categories. Due to sensitivity shown by the maximum t-statistic approach in simulations, it is possible that we are getting signals from non-topological difference between the two groups of power maps.

5 Discussion

In this paper, the spatial distribution of an EEG topographic power map is quantified through a novel node-based network filtration. We use the network filtration to compare the spatial distribution of EEG power maps in long-term meditators and meditation naïve practitioners. The results show that the meditators have on average fewer clusters, thus a more coherent spatial distribution, than novices in the early stage of NREM sleep.

In EEG analysis, a general concern is that the scalp signal at each electrode is a weighted sum of the signal generated by all cortical sources. For future research, we will also explore an unmixing procedure such as working in source space after applying a distributed solution and analyzing selected independent components. It will provide deeper insight into the underlying neurophysiological dynamics that the topological network analysis has the potential to capture.

Acknowledgment. This work was supported by the National Center for Complementary and Alternative Medicine (NCCAM) P01AT004952. We also acknowledge the support of NIH grants UL1TR000427 and EB022856.

References

1. Brunet, D., Murray, M.M., Michel, C.M.: Spatiotemporal analysis of multichannel EEG: CARTOOL. Comput. Intell. Neurosci. **2011**, 15 (2011)
2. Davidson, R.J., Lutz, A.: Buddha's brain: neuroplasticity and meditation. IEEE Signal Process. Mag. **25**(1), 176–174 (2008)
3. Dentico, D., Ferrarelli, F., Riedner, B.A., Smith, R., Zennig, C., Lutz, A., Tononi, G., Davidson, R.J.: Short meditation trainings enhance non-REM sleep low-frequency oscillations. PLoS ONE **11**(2), e0148961 (2016)
4. Edelsbrunner, H., Harer, J.: Computational Topology. American Mathematical Society, Providence (2010)
5. Eklund, A., Nichols, T.E.T.E., Knutsson, H.: Cluster failure: why fMRI inferences for spatial extent have inflated false-positive rates. Proc. Nat. Acad. Sci. **113**(33), 7900–7905 (2016)
6. Ferrarelli, F., Smith, R., Dentico, D., Riedner, B.A., Zennig, C., Benca, R.M., Lutz, A., Davidson, R.J., Tononi, G.: Experienced mindfulness meditators exhibit higher parietal-occipital EEG gamma activity during NREM sleep. PLoS ONE **8**(8), e73417 (2013)

7. Maris, E.: Statistical testing in electrophysiological studies. Psychophysiology **49**(4), 549–565 (2012)

8. Maris, E., Oostenveld, R.: Nonparametric tatistical testing of EEG- and MEG-data. J. Neurosci. Methods **164**(1), 177–90 (2007)

9. Mensen, A., Khatami, R.: Advanced EEG analysis using threshold-free cluster-enhancement and non-parametric statistics. NeuroImage **67**, 111–118 (2013)

10. Welch, P.: The use of fast Fourier transform for the estimation of power spectra: A method based on time averaging over short, modified periodograms. IEEE Trans. Audio Electroacoust. **15**(2), 70–73 (1967)

Region-Wise Stochastic Pattern Modeling for Autism Spectrum Disorder Identification and Temporal Dynamics Analysis

Eunji Jun and Heung-Il Suk[✉]

Department of Brain and Cognitive Engineering,
Korea University, Seoul, Republic of Korea
hisuk@korea.ac.kr

Abstract. Many studies in the literature have validated the use of resting-state fMRI (rs-fMRI) for brain disorder/disease identification. Unlike the existing methods that mostly first estimate functional connectivity and then extract features with a graph theory, in this paper, we propose a novel method that directly models the temporal stochastic patterns inherent in BOLD signals for each Region Of Interest (ROI) individually. Specifically, we model temporal BOLD signal fluctuation of an individual ROI by means of Hidden Markov Models (HMMs), and then compute a regional BOLD signal likelihood with the trained HMMs. By regarding the BOLD signal likelihood of ROIs over a whole brain as features, we build a classifier that can discriminate subjects with Autism Spectrum Disorder (ASD) from Normal healthy Controls (NC). In addition, we also devise a method to further investigate the characteristics of temporal dynamics in rs-fMRI estimated by HMMs. For group comparison, we use the metrics of state occupancy rate and lifetime of the optimal hidden states that best represent the temporal BOLD signals. In our experiments with ABIDE cohort, we validated the effectiveness of the proposed method by achieving the highest diagnostic accuracies among competing methods. We could also identify the group differences in temporal dynamics between ASD and NC in terms of state occupancy rate and lifetime of individual states.

1 Introduction

Recent studies have witnessed that functional connectivity during resting, *i.e.*, not performing any cognitive task, is changing over time [2]. Since those findings, various attempts have been made to effectively investigate intrinsic functional dynamics of the brain network [9, 11, 16, 17]. However, many existing studies considering brain dynamics depend mostly on a sliding window approach [1]. A fixed length of window is shifted over time resulting in time-varying covariance matrices between brain regions for each sliding window and the measures are clustered to construct intrinsic brain networks [10, 12]. But this approach has a limitation that the size of the sliding window and the number of clusters are

© Springer International Publishing AG 2017
G. Wu et al. (Eds.): CNI 2017, LNCS 10511, pp. 143–151, 2017.
DOI: 10.1007/978-3-319-67159-8_17

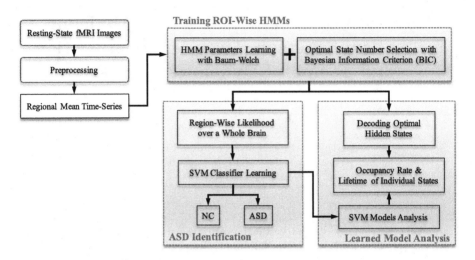

Fig. 1. An overview of the proposed framework for ASD identification with rs-fMRI and analysis of the trained model.

often chosen in an arbitrary manner, which may lead to inaccurate estimation of the underlying functional dynamics.

An alternative approach, which we also utilize in this work, is based on the general framework of probabilistic generative models, which enable quantitative analysis about key properties of temporal BOLD fluctuations. To assess temporal dynamics, we use Hidden Markov Models (HMMs) [14], which assume that observations are combined through a set of time-dependent hidden-state variables represented as a first-order Markov chain. Thus, unlike the sliding window approach that requires exact window length, HMMs systematically evaluate long-term dependencies by modeling transitions among states, which can be inferred from observation stochastically. Earlier, Eavani *et al.* [6] used HMMs to infer dynamic functional connectivities based on temporal BOLD signals by explicitly introducing sparse inverse covariance matrices in an HMM. Despite their sound modeling for dynamic functional connectivity estimation, due to the large number of parameters, mostly involved in inverse covariance matrices, that should be learned from samples, it showed very limited performance in brain disease diagnosis [17].

In this work, we propose a novel framework of modeling temporal dynamics inherent in rs-fMRI with HMMs, based on which we further devise a method of discriminating Autism Spectrum Disorder (ASD) from Normal healthy Controls (NC). The framework of the proposed method is schematized in Fig. 1. To overcome the limitation of previous application of HMMs in rs-fMRI, we explicitly model dynamic characteristics of ROIs from regional mean time series of rs-fMRI, which we thus call as an '*ROI-Wise HMMs*'. Based on the trained HMMs, we then compute '*goodness-of-fit*' of regional mean time-series, *i.e.*, likelihood, from the corresponding HMM. We define a whole-brain feature vector by concatenating regional likelihoods, which is then fed into a classifier for ASD diagnosis. Along with the ASD diagnosis, motivated by significant evidences in recent study [19] that patients with ASD exhibit different temporal dynamics, we also present a novel method of estimating temporal dynamics in terms of state occupancy rate and lifetime of individual states in the whole brain as a way to evaluate group differences.

2   Materials and Image Processing

We acquired preprocessed rs-fMRI data from the University of Michigan (UM) open to the public in the Autism Brain Imaging Data Exchange (ABIDE)[1] [5]. The samples are 120 subjects in total, including 47 ASD, and 73 NC. During preprocessing, the first four volumes of each subject were initially discarded prior to further processing to ensure magnetization equilibrium. The remaining volumes were then spatially normalized to the MNI space with a voxel size of $3 \times 3 \times 3 \, \text{mm}^3$. Regression of nuisance signals including ventricle, white matter, and global signals was performed by applying Friston 24-parameter model [8]. The regressed rs-fMRI images were parcellated into 116 ROIs using the Automated Anatomical Labeling (AAL) atlas [18], and the mean time series of the BOLD signals in voxels of each ROI were computed. The mean signals were band-pass filtered from 0.01 to 0.1 Hz to exploit the characteristics of low frequency fluctuation in rs-fMRI, resulting in 116-dimensional vectors for each subject.

3   ROI-Wise Temporal Dynamics Estimation

3.1   ROI-Wise HMMs Modeling

Assume that we are given a sequence of T-length mean time-series of rs-fMRI from the r-th ROI, $i.e.$, $\mathbf{X}^r = \left[(\mathbf{x}_1^r)^\top; \cdots ; (\mathbf{x}_n^r)^\top; \cdots ; (\mathbf{x}_N^r)^\top \right] \in \mathbb{R}^{N \times T}$, where $r = \{1, ..., R\}$, $\mathbf{x}_n^r = \left[x_{n,1}^r, \cdots, x_{n,t}^r, \cdots, x_{n,T}^r \right]^\top \in \mathbb{R}^T$, R and N denote, respectively, the number of ROIs and the number of subjects. We hypothesize that (i) the temporal BOLD fluctuations of ROIs has their own stochastic patterns and (ii) such temporal stochastic patterns between NC and ASD may be different, which could thus be useful to exploit temporal stochastic patterns for ASD and NC discrimination. Based on these hypotheses, we model region-wise temporal dynamics with HMMs individually, which we call as *ROI-Wise HMMs*.

Let s_t^r and o_t^r denote random variables of a hidden state and an observation for an HMM of an ROI r at time t, respectively. With a first-order Markov chain assumption, the hidden state s_t^r at time t is dependent on the hidden state s_{t-1}^r

[1] ABIDE (http://fcon_1000.projects.nitrc.org/indi/abide) provides preprocessed rs-fMRI datasets for ASD and NC by performing four different preprocessing pipelines. In this work, we used datasets preprocessed by the Data Processing Assistant for Resting-State fMRI (DPARSF), a convenient plug-in software based on SPM and REST.

at time $(t-1)$. Meanwhile, the observable variable o_t^r at time t is dependent on the hidden state s_t^r at the same time. Then, in HMM, there are two underlying stochastic processes, namely, a hidden state process $P\left(s_t^r|s_{t-1}^r\right)$ and an observation process $p\left(o_t^r|s_t^r\right)$ over $t \in \{1,\ldots,T\}$. When we consider a K number of hidden states, the unobservable inherent dynamics in rs-fMRI is modeled by the hidden state process, *i.e.*, a state transition probability $A^r = \left[a_{ij}^r\right]_{i,j\in\{1,\ldots,K\}}$, where $a_{ij}^r \equiv P\left(s_t^r = j|s_{t-1}^r = i\right)$ denotes the probability of changing from one hidden state $(s_{t-1}^r = i)$ to another hidden state $(s_t^r = j)$, and an initial state probability $\Pi^r = [\pi_i^r]_{i\in\{1,\ldots,K\}}$, where $\pi_i^r \equiv P\left(s_1^r = i\right)$ denotes the probability of starting from a specific hidden state at time $t = 1$. We define an ergodic topology for the state transition, which allows the temporal BOLD fluctuation to change from one state to any other. The observable process is represented by an emission probability density function (*pdf*) $B^r = \{b_i^r\}_{i\in\{1,\ldots,K\}}$, where $b_i^r \equiv p\left(o_t^r = x_{n,t}^r|s_t^r = i\right)$ denotes the likelihood of observing the specific observation $x_{n,t}^r$ when staying at the hidden state of i. As for an emission pdf b_i^r, a univariate Gaussian distribution is used. Thus, we completely define an ROI-Wise HMM with the parameter set of $\lambda^r = (\Pi^r, A^r, B^r)$.

With regard to the number of hidden states, K, we allow different ROIs to have different number of states. In our experiments, we utilized Bayesian Information Criterion (BIC) [20] to select the optimal value of K^r for each ROI. By training ROI-Wise HMMs with a Baum-Welch algorithm [14], we use them as a way to represent the goodness-of-fit for the given mean time-series of the corresponding ROI.

3.2 Feature Extraction and Classifier Learning

Given a mean time-series of the r-th ROI for a subject n, *i.e.*, $\mathbf{x}_n^r$, we compute how likely the sequence $\mathbf{x}_n^r$ is generated from the corresponding trained HMM $\hat{\lambda}^r$ as follows:

$$p\left(\mathbf{x}_n^r|\hat{\lambda}^r\right) = \sum_{\mathbf{s}_n^r} p\left(\mathbf{x}_n^r|\mathbf{s}_n^r, \hat{\lambda}^r\right) P\left(\mathbf{s}_n^r|\hat{\lambda}^r\right) \tag{1}$$

where $\hat{\mathbf{s}}_n^r = \left[\hat{s}_{n,1}^r, \cdots, \hat{s}_{n,t}^r, \cdots, \hat{s}_{n,T}^r\right]^\top$ denotes a sequence of the hidden states. Equation (1), which can be efficiently computed by forward algorithm [14], basically calculates the *goodness-of-fit* of the HMM $\hat{\lambda}^r$ to represent the observation $\mathbf{x}_n^r$. Note that the HMM $\hat{\lambda}^r$ is learned by taking into account the mean time-series of the r-th ROI over all subjects in a training dataset. Hence, for those who have different temporal patterns in their mean time-series, it is expected to have low likelihoods obtained from the corresponding model.

In this paper, we utilize the regional goodness-of-fit over a whole brain of a subject n, computed with the respective ROI-wise HMMs, as features:

$$\mathbf{f}_n = \left[p\left(\mathbf{x}_n^r|\hat{\lambda}^r\right)\right]_{r=1}^R \in \mathbb{R}^R. \tag{2}$$

The feature vector $\mathbf{f}_n$ can be also regarded as a likelihood map, representing the regional abnormality over a whole-brain in terms of temporal dynamics.

For ASD identification, we use a linear Support Vector Machine (SVM), one of the most widely used classifiers for brain disease diagnosis in the literature [3,4,7]. In particular, we consider two different SVM models, namely, standard SVM and an ℓ_1-norm SVM. The ℓ_1-norm SVM has the effect of automatically selecting class-discriminative features by explicitly introducing a ℓ_1-penalty term over the parameter values in the loss function as follows:

$$\min_{\beta_0,\beta} \sum_{i=1}^{n} \left[1 - y_i \left(\beta_0 + \sum_{j=1}^{R} \beta_j h_j(x_i) \right) \right]_+ \quad s.t. \ \|\beta\|_1 = |\beta_1| + \cdots + |\beta_q| \le \gamma \quad (3)$$

where $[\cdot]_+$ denotes a hinge loss function, $y_i \in \{+1, -1\}$ indicates a class label of the i-th sample, γ is a tunable hyperparameter, and $h(x)$ is a basis function, for which we use an identity function in this paper.

3.3 Measuring Temporal Dynamics

We also propose a method of quantifying the temporal dynamics probabilistically from the trained HMMs as a way of identifying group differences. We first decode the optimal state sequence of an observation with a Viterbi algorithm [14], where each hidden node produces the maximum likelihood of the observation sequence from time 1 to T. Based on these decoding results, we provide quantitative information of characterizing the temporal dynamics over BOLD signals in a brain. Specifically, inspired by the work of [15], we consider occupancy rate and lifetime of hidden states, with which we attempt to find out the group differences in region-wise temporal dynamics.

The occupancy rate of the hidden state i for a subject n is defined as follows:

$$O(n,i) = \frac{1}{T} \sum_{t=1}^{T} \delta(s_{n,t}, i) \times 100(\%) \quad (4)$$

where $\delta(s_{n,t}, i)$ is the Dirac delta function, whose value is one if the current state at time t is equal to i or zero otherwise. In the mean time, the mean lifetime of a state is defined as the average time that a given state i continues to last before transitioning to another state.

4 Experimental Settings and Results

4.1 Choosing Optimal Number of States in HMMs

To build an ROI-wise HMM, we used mean time-series of a respective ROI from all subjects in our dataset in an unsupervised manner. In order to choose the number of states K^r for an HMM of the r-th ROI, we used the BIC, which is defined for our case as follows:

$$BIC(\lambda^r) = \log P(\mathbf{X}^r; \lambda^r) - \frac{D}{2} \log N \quad (5)$$

where $P(\mathbf{X}^r; \lambda^r)$ denotes a marginal likelihood of the training data $\mathbf{X}^r$, which represents how well the model λ^r represents the data, and D and N denote, respectively, the number of parameters in λ^r and the number of training samples in $\mathbf{X}^r$. As for the K^r, we varied the values in $\{1, \ldots, 25\}$ and selected the one whose BIC score is the largest. The statistics of the number of states chosen by BIC across all ROIs were 23(mean) ± 2.3(std) and 15(min)-25(max).

4.2   Performance Comparison

In regard to performance evaluation, we took a 10-fold cross-validation technique. Specifically, we partitioned samples of each class, $i.e.$, ASD and NC, into 10 folds and used samples of one fold for test and those of the remaining folds for training. We repeated the process 10 times and reported the average of the results as performance below.

To validate the effectiveness of the proposed method, we compared our method with two existing methods in the literature, which estimate functional connectivity with (i) Pearson correlation [13], and (ii) sliding window-based Dynamic Functional Network (sDFN) [10]. For both methods, we extracted weighted clustering coefficients based on a graph theory as features. All the competing methods commonly used standard and ℓ_1-norm SVM for classification. Concisely, a linear kernel was used with the model parameter C chosen in the set of $\{10^{-5}, 10^{-4}, \ldots, 10^4\}$ by nested cross-validation.

Table 1 shows the performance of the competing methods. It is remarkable that the proposed method outperformed the other methods by 35.02% (vs. Pearson), and 25.31% (vs. sDFN) with standard SVM and 30.91% (vs. Pearson), and 25.8% (vs. sDFN) with ℓ_1-norm SVM. It is also noteworthy that the proposed method achieved the AUC of 0.97 and 0.94 with standard SVM and ℓ_1-norm SVM, respectively, suggesting the great potential for practical use in the clinic.

Table 1. Performance comparison with the competing methods by standard and ℓ_1-norm SVM. (AUC: Area Under the receiver operator characteristic Curve)

	Methods	Accuracy (%)	Sensitivity (%)	Specificity (%)	AUC
Standard SVM	Pearson [13]	55.07 ± 22.32	48.50	59.29	0.58
	sDFN [10]	64.78 ± 6.12	22.50	91.79	0.65
	Ours	$\mathbf{90.09 \pm 6.51}$	**85.50**	**93.21**	**0.97**
ℓ_1-norm SVM	Pearson [13]	57.58 ± 12.78	62.50	55.00	0.60
	sDFN [10]	62.69 ± 12.62	66.00	60.54	0.64
	Ours	$\mathbf{88.49 \pm 9.02}$	**87.50**	**89.29**	**0.94**

4.3   Regional Importance and Temporal Dynamics Analysis

For regional importance analysis in discriminating between ASD and NC, we first calculated the statistical significance of the features, $i.e.$, goodness-of-fit or

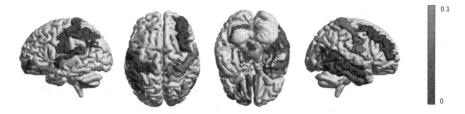

Fig. 2. Statistical map of p-values obtained by conducting permutation tests between ASD and NC with goodness-of-fit features. For unclustered, regions of low p-values (smaller than 0.1) are only marked.

likelihood, by means of permutation tests between ASD and NC. The resulting p-value map projected into a brain is shown in Fig. 2, where regions with p-value smaller than 0.1 are marked only for uncluttered.

While the proposed method showed its superiority in discriminating ASD from NC in Sect. 4.2, it doesn't provide any insight into the temporal dynamic characteristics inherent in rs-fMRI. In this regard, we further conducted to analyze the learned HMMs by focusing on those ROIs, *i.e.*, marked in Fig. 2. Specifically, we obtained a single state sequence that best represents the observation sequence by Viterbi algorithm. Based on the decoded state sequences, we computed occupancy rate and lifetime of individual states described in Sect. 3.3. Figure 3 shows their histograms for ASD and NC, respectively. Occupancy rate varies between 5.77% and 18.72% with average of $5.00 \pm 3.12\%$ in ASD, and between 5.57% and 16.53% with average of $5.00 \pm 2.40\%$ in NC. Meanwhile, mean lifetime varies between 0.14 s and 2.84 s, with average of 1.13 ± 0.52 s in ASD, and between 0.15 s and 2.88 s with average of 1.23 ± 0.41 s in NC. Therefore, we could identify different patterns between ASD and NC in the dynamic characteristics.

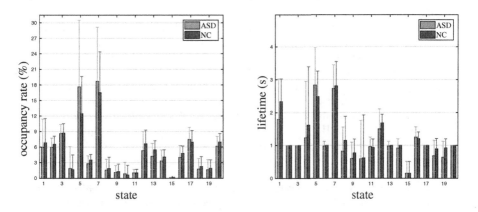

Fig. 3. Occupancy rate (left) and mean lifetime (right) of the decoded state sequences

5 Conclusion

In this paper, we proposed a novel framework for rs-fMRI based ASD diagnosis and temporal dynamics analysis. Specifically, we exploited likelihoods of a BOLD sequence computed from ROI-wise HMMs as features by hypothesizing that the dynamic patterns of BOLD signal may be different between ASD and NC. Experimental results showed that the proposed method achieved higher performance than the competing methods considered in this work. In addition, we found the statistically significant ROIs and identified the group differences in *temporal dynamics* between ASD and NC by means of quantitative analysis with the decoded hidden state sequences.

Acknowledgement. This research was supported by Basic Science Research Program through the National Research Foundation of Korea (NRF) funded by the Ministry of Education (NRF-2015R1C1A1A01052216) and also partially supported by Institute for Information & Communications Technology Promotion (IITP) grant funded by the Korea government (No. 2017-0-00451).

References

1. Allen, E.A., Damaraju, E., Plis, S.M., Erhardt, E.B., Eichele, T., Calhoun, V.D.: Tracking whole-brain connectivity dynamics in the resting state. Cerebral Cortex p. bhs352 (2012)
2. Chang, C., Glover, G.H.: Time-frequency dynamics of resting-state brain connectivity measured with fMRI. NeuroImage **50**(1), 81–98 (2010)
3. Chen, H., Duan, X., Liu, F., Lu, F., Ma, X., Zhang, Y., Uddin, L.Q., Chen, H.: Multivariate classification of autism spectrum disorder using frequency-specific resting-state functional connectivity: a multi-center study. Prog. Neuropsychopharmacol. Biol. Psychiatry **64**, 1–9 (2016)
4. Craddock, R.C., Holtzheimer, P.E., Hu, X.P., Mayberg, H.S.: Disease state prediction from resting state functional connectivity. Magn. Reson. Med. **62**(6), 1619–1628 (2009)
5. Di Martino, A., Yan, C.G., Li, Q., Denio, E., Castellanos, F.X., Alaerts, K., Anderson, J.S., Assaf, M., Bookheimer, S.Y., Dapretto, M., et al.: The autism brain imaging data exchange: towards a large-scale evaluation of the intrinsic brain architecture in autism. Mol. Psychiatry **19**(6), 659–667 (2014)
6. Eavani, H., Satterthwaite, T.D., Gur, R.E., Gur, R.C., Davatzikos, C.: Unsupervised learning of functional network dynamics in resting state fMRI. In: Information Processing in Medical Imaging, vol. 23, p. 426. NIH Public Access (2013)
7. Fan, Y., Liu, Y., Wu, H., Hao, Y., Liu, H., Liu, Z., Jiang, T.: Discriminant analysis of functional connectivity patterns on Grassmann manifold. NeuroImage **56**(4), 2058–2067 (2011)
8. Friston, K.J., Williams, S., Howard, R., Frackowiak, R.S., Turner, R.: Movement-related effects in fMRI time-series. Magn. Reson. Med. **35**(3), 346–355 (1996)
9. Gilbert, C.D., Sigman, M.: Brain states: top-down influences in sensory processing. Neuron **54**(5), 677–696 (2007)
10. Leonardi, N., Van De Ville, D.: On spurious and real fluctuations of dynamic functional connectivity during rest. NeuroImage **104**, 430–436 (2015)

11. Li, X., Lim, C., Li, K., Guo, L., Liu, T.: Detecting brain state changes via fiber-centered functional connectivity analysis. Neuroinformatics **11**(2), 193–210 (2013)
12. Lindquist, M.A., Xu, Y., Nebel, M.B., Caffo, B.S.: Evaluating dynamic bivariate correlations in resting-state fMRI: a comparison study and a new approach. NeuroImage **101**, 531–546 (2014)
13. Nielsen, J.A., Zielinski, B.A., Fletcher, P.T., Alexander, A.L., Lange, N., Bigler, E.D., Lainhart, J.E., Anderson, J.S.: Multisite functional connectivity MRI classification of autism: Abide results. Front. Hum. Neurosci. **7**, 599 (2013)
14. Rabiner, L.R.: A tutorial on hidden Markov models and selected applications in speech recognition. Proc. IEEE **77**(2), 257–286 (1989)
15. Ryali, S., Supekar, K., Chen, T., Kochalka, J., Cai, W., Nicholas, J., Padmanabhan, A., Menon, V.: Temporal dynamics and developmental maturation of salience, default and central-executive network interactions revealed by variational bayes hidden markov modeling. PLoS Comput. Biol. **12**(12), e1005138 (2016)
16. Smith, S.M., Miller, K.L., Moeller, S., Xu, J., Auerbach, E.J., Woolrich, M.W., Beckmann, C.F., Jenkinson, M., Andersson, J., Glasser, M.F., et al.: Temporally-independent functional modes of spontaneous brain activity. Proc. Natl. Acad. Sci. **109**(8), 3131–3136 (2012)
17. Suk, H.I., Wee, C.Y., Lee, S.W., Shen, D.: State-space model with deep learning for functional dynamics estimation in resting-state fMRI. NeuroImage **129**(1), 292–307 (2016)
18. Tzourio-Mazoyer, N., Landeau, B., Papathanassiou, D., Crivello, F., Etard, O., Delcroix, N., Mazoyer, B., Joliot, M.: Automated anatomical labeling of activations in SPM using a macroscopic anatomical parcellation of the MNI MRI single-subject brain. NeuroImage **15**(1), 273–289 (2002)
19. Washington, S.D., Gordon, E.M., Brar, J., Warburton, S., Sawyer, A.T., Wolfe, A., Mease-Ference, E.R., Girton, L., Hailu, A., Mbwana, J., et al.: Dysmaturation of the default mode network in autism. Hum. Brain Mapp. **35**(4), 1284–1296 (2014)
20. Bishop, C.M.: Pattern recognition. Mach. Learn. **128**, 1–58 (2006)

A Whole-Brain Reconstruction Approach for FOD Modeling from Multi-Shell Diffusion MRI

Wei Sun, Junling Li, and Yonggang Shi[✉]

Laboratory of Neuro Imaging, USC Stevens Neuroimaging and Informatics Institute,
Keck School of Medicine, University of Southern California, Los Angeles, USA
yonggang.shi@loni.usc.edu

Abstract. With the advance of connectome imaging techniques, there is a great need of robust methods for modeling the distribution of fiber orientations from multi-shell diffusion imaging. Existing tools for fiber orientation distribution (FOD) reconstruction, however, predominantly solves this problem on a voxel-by-voxel basis, disregarding the spatial regularity in brain anatomy. In this work, we propose a novel computational framework for the joint reconstruction of FODs over the whole brain volume. Our framework takes into account compartment modeling from multi-shell imaging data and uses an operator splitting scheme to decouple the whole-brain reconstruction problem into a series of local computations. Within this framework, we can investigate both isotropic and anisotropic regularizations. In the experiments, we conduct extensive simulations to compare the performance of both types of regularizations and show that anisotropic regularization produces more robust results across various fiber configurations. We also apply our method to *in vivo* data from 80 HCP subjects and evaluate the impact of FOD modeling methods on the reconstruction of the challenging fiber bundles from the locus coeruleus (LC) nuclei. Our results indicate that the proposed whole-brain approach for FOD modeling leads to more robust LC fiber bundle reconstruction than results from voxel-wise modeling.

1 Introduction

With the success of the Human Connectome Project (HCP) [1], cutting-edge multi-shell imaging is becoming increasingly popular [2], which has motivated the development of novel algorithms for processing these sophisticated imaging data. In particular, the multi-shell imaging data has enabled the joint estimation of FOD and compartment parameters [3–5]. Popular FOD reconstruction methods, however, remains tackling this problem in a voxel-wise fashion while ignoring the 3D geometry and regularity across the human brain. In this work, we develop

---

Y. Shi—This work was in part supported by the National Institute of Health (NIH) under Grant R01EB022744, P41EB015922, U01EY025864, U01AG051218, P50AG05142.

© Springer International Publishing AG 2017
G. Wu et al. (Eds.): CNI 2017, LNCS 10511, pp. 152–160, 2017.
DOI: 10.1007/978-3-319-67159-8_18

a whole-brain approach for FOD reconstruction from multi-shell imaging data by incorporating both compartment modeling and spatial regularity.

Several previous methods based on either post-processing or regularized reconstruction were proposed to introduce spatial regularization into the estimation of FODs. After the voxel-wise reconstruction of FODs, smoothing can be applied across the 3D space to improve regularity [6,7]. During the FOD reconstruction stage, isotropic regularization of the spherical harmonics (SPHARM) coefficients of FODs were used in [8,9] for improving the smoothness of FOD fields. An orientation-specific, fiber continuity (FC) model was introduced in [10] to selectively penalizing FOD differences along different angles. Overall these previous methods were developed in single-shell imaging and have been evaluated on mostly very small number (typically 1) of subjects.

In this work, we propose a novel, whole-brain FOD reconstruction method from multi-shell imaging data. At each voxel, we leverage a multi-compartment model proposed recently [5] to characterize the multi-shell imaging signal. For the incorporation of spatial regularization, we develop an iterative algorithm motivated by the operator splitting schemes in energy minimization [11]. Our computational framework is also flexible and can utilize both isotropic and anisotropic regularization. In our experiments, we conduct extensive simulations to compare the performance of voxel-wise reconstruction and whole-brain reconstruction with these two types of regularizations. We demonstrate that anisotropic regularization achieves more robust performance in different fiber configurations. After that, we apply our method to *in vivo* data from 80 HCP subjects for the reconstruction of the challenging bundles from the locus coeruleus, which is emerging as a critical area of Alzheimer's disease research [12]. We show that FODs computed by our whole-brain reconstruction method leads to more robust fiber bundle extraction than FODs from voxel-wise reconstruction.

2   Method

2.1   Voxel-Wise FOD Modeling from Multi-Shell Imaging

Let Ω denote the image volume. Following [3,5], we model each voxel $\mathbf{p} \in \Omega$ as composed of three compartments that contribute to the diffusion imaging signal: intra-axonal compartment, extra-axonal compartment and the DOT model with negligible diffusion. Let $\bar{s}$ denote a vector of signals from N gradient directions distributed over multiple b-values. We can define its compartmental formulation as:

$$\bar{s} = A\bar{x} + \alpha\bar{\beta} + \gamma\bar{e} + \bar{n}. \tag{1}$$

For the intra-axonal compartment, the matrix A represents the spherical convolution of the FOD with a stick kernel, where $\bar{x}$ is the vector of SPHARM coefficients for the FOD. The second term on the right side models the extra-axonal compartment with isotropic diffusion, where $\bar{\beta} = [e^{-b_1\lambda_{iso}} e^{-b_2\lambda_{iso}} \cdots e^{-b_N\lambda_{iso}}]^T$. Both the diffusivity λ_{iso} and volume fraction α are unknown and should be estimated from the data. In the third term, γ is the volume fraction of the DOT model to be estimated, and $\bar{e} = [11 \cdots 1]^T$ is a vector of length N. Finally, $\bar{n}$ is the vector of noise. For voxel-wise reconstruction, we can solve the following constrained energy minimization problem [5]:

$$\text{minimize} \quad E(\bar{x}, \alpha, \lambda_{iso}, \gamma) = \frac{1}{2} \left\| \bar{s} - [A \, \bar{\beta} \, \bar{e}] \begin{bmatrix} \bar{x} \\ \alpha \\ \gamma \end{bmatrix} \right\|^2 + \xi I \bar{x} \tag{2}$$

subject to $I\bar{x} + \alpha + \gamma = 1$, $C_M \bar{x} \geq 0$, $\alpha \geq 0$, $\lambda_{iso} \geq 0$, and $\gamma \geq 0$.

The first term in the energy is a data fidelity term, the second term is a sparsity penalty term for FOD, and ξ is the weighting factor of sparsity term. The constraints include the normalization condition of the volume fractions of the three compartments, and non-negativity of the FOD and the volume fractions α and γ. More specificially, $I\bar{x}$ denotes the volume fraction of the intra-axonal compartment with $I = [\sqrt{4\pi} \ 0 \cdots 0]$. The matrix C_M represents an adaptively selected constraint matrix to guarantee the non-negativity of FOD.

2.2   Whole-Brain FOD Modeling with Spatial Regularity

For the joint reconstruction of the FODs over the whole image volume, we define an energy function with spatial regularization as:

$$E^* = \sum_{\mathbf{p} \in \Omega} \left(E_{\mathbf{p}} + \frac{\omega}{2} \|\nabla v(\mathbf{p})\|^2 \right) \tag{3}$$

where $E_{\mathbf{p}}$ denote the energy in (2) at each voxel $\mathbf{p}$, $v(\mathbf{p}) = [\bar{x}(\mathbf{p}) \, \alpha(\mathbf{p}) \, \gamma(\mathbf{p})]^T$, ω is a weighting factor, and ∇ is a gradient operator for each element of v over Ω. The gradient operator couples FODs over neighboring voxels and henceforth encourages their smoothness spatially for more robust reconstruction over the whole image volume.

To solve this high dimensional optimization problem on the order of 10^8 variables (around 1 million voxels times approximately 100 SPHARM coefficients), we develop an iterative algorithm based on operator splitting. Let v_k denote the solution of v at the k-th iteration. In the first step of each iteration, we compute an auxiliary variable v_k^s from the current solution of v_k via a Laplacian diffusion:

$$v_k^s(\mathbf{p}) = v_k + \tau \Delta v_k(\mathbf{p}), \tag{4}$$

where $k \in [0, K]$ represents the iteration number, the auxiliary variable $v_k^s(\mathbf{p})$ denote the value at the voxel $\mathbf{p} \in \Omega$, τ is the Gaussian diffusion duration and Δ represents the Laplace operator applied to each element of v_k independently. In the second step, we minimize the auxiliary energy function Eq. (5) <u>at each voxel</u> using

$$\begin{aligned} E_{k+1}^*(\mathbf{p}) = {} & E_{k+1}(\mathbf{p}) + \frac{\omega}{2\tau} \|v_{k+1}(\mathbf{p}) - v_k^s(\mathbf{p})\|^2, \\ = {} & \frac{1}{2} \left\| \bar{s}(\mathbf{p}) - [A \, \bar{\beta} \, \bar{e}] \, v_{k+1}(\mathbf{p}) \right\|^2 + \xi I \bar{x}(\mathbf{p}) + \frac{\omega}{2\tau} \|v_{k+1}(\mathbf{p}) - v_k^s(\mathbf{p})\|^2. \end{aligned} \tag{5}$$

The last term in Eq. (5) decouples the computational complexity of FOD estimation and spatial regularization. Note that its weight is inversely proportional to the diffusion time τ in (4). This is intuitively easy to understand that less weight should be put on the auxiliary term if it is overly smoothed. In contrast to Eq. (3), Eq. (5) is a voxel-wise operation because the spatial regularization has been split into Eq. (4). Thus the computational cost is expected to only increase linearly with respect to the number of iterations as compared to voxel-wise reconstruction.

2.3   Spatial Regularization via Hyper-spherical Smoothing

We describe next the numerical implementation of the diffusion operation in Eq. (4) on the FODs and compartment parameters. For the smoothing of FODs, we utilize a hyper-spherical smoothing technique to preserve their energy and avoid shrinkage. Let $\bar{x}_{nc}$ and $\bar{x}_{nn}$ denote the normalized coefficients of SPHARMs located at a center voxel and one of its neighbor voxels, respectively. We first project $\bar{x}_{nn}$ to the tangent plane at $\bar{x}_{nc}$ as

$$\bar{x}'_{nn} = r\bar{x}_{nn} - (1 - r)\bar{x}_{nc}, \tag{6}$$

where scalar $r = 2/(\bar{x}_{nc} \cdot \bar{x}_{nn} + 1)$. On this tangent plane, the diffusion of $\bar{x}_{nc}$ is:

$$\bar{x}'_{nc} = \bar{x}_{nc} + \tau \sum_{i=1}^{4} g(\|\bar{x}'_{nni} - \bar{x}_{nc}\|)(\bar{x}'_{nni} - \bar{x}_{nc}). \tag{7}$$

where $\bar{x}'_{nni}$ denotes the stereographic projection from the i-th neighbor ($i = 1, \cdots, 4$). The function $g(\|\cdot\|) = 1$ for isotropic diffusion and $g(\|\cdot\|) = e^{-(\|\cdot\|/\sigma)^2}$ for anisotropic diffusion. The parameter σ controls the effect range of anisotropic smoothing, and $\|\cdot\|$ is the L_2-norm.

After the smoothing process, $\bar{x}'_{nc}$ should be mapped backward to the sphere surface. The backward mapping is expressed as:

$$\bar{x}^s_{nc} = r'\bar{x}'_{nc} - (1 - r')\bar{x}_{nc}, \tag{8}$$

where scalar $r' = 2(\bar{x}'_{nc} \cdot \bar{x}_{nc} + 1)/(\bar{x}'_{nc} \cdot \bar{x}'_{nc} + 2\bar{x}'_{nc} \cdot \bar{x}_{nc} + 1)$, and $\bar{x}^s_{nc}$ stands for smoothed $\bar{x}_{nc}$ using spherical smoothing. The original magnitude of $\bar{x}_c$ should be multiplied back to recover the magnitude of $\bar{x}^s_c$. Through spherical smoothing we can preserve the energy of original FOD and avoid shrinkage.

For the scalar compartment fractions α and γ in v_k, we apply similar isotropic or anisotropic smoothing using Eq. (7) but without spherical projection.

3   Experiments

In this section, we present experimental results to demonstrate the proposed whole-brain FOD reconstruction method on simulated and *in vivo* multi-shell imaging data from HCP. Three candidate approaches were compared: the baseline method of voxel-wise reconstruction in [5] without spatial regularization; the proposed reconstruction method with isotropic or anisotropic regularizations. We fixed the following parameters for all experiments on simulation and real data. For the energy in Eqs. (2) and (5), we fixed weighting factor of sparsity term ξ to 0.2. In Eq. (7), Gaussian smoothing duration τ controls the convergence and stability of the algorithm. For both isotropic and anisotropic diffusion, the largest allowable value is selected for τ that still ensures numerical stability. More specifically, τ is chosen as 0.1 and 0.25 for isotropic and anisotropic regularizations, respectively. For anisotropic regularization, σ is fixed to 0.5. We set the weighting coefficient of spatial regularization term $\omega = 20$. For the iteration numbers of algorithms, we set $K = 20$ on simulated data and $K = 10$ on HCP data.

3.1 Simulation

We followed the LifeSpan protocol of HCP that consists of 98 gradient directions distributed over two b-values $1500s/mm^2$ and $3000s/mm^2$ to generate the simulated data. The simulated intra-axonal, extra-axonal, and trapped water fractions were fixed as: 0.35, 0.5, and 0.15. The diffusivity of the extra-axonal compartment is fixed as $0.0012\ mm^2/s$ in the simulation. Two fiber patterns are used in our simulation experiments. In the first experiment, we use the pattern in Fig. 1(A) to simulate the multi-shell imaging data and evaluate the performance of FOD reconstruction algorithms when there are sharp turning angles between

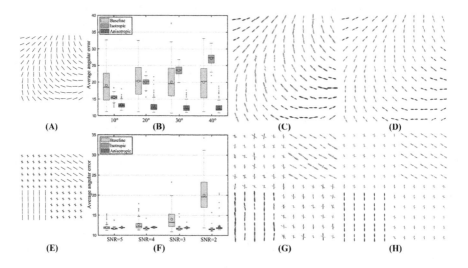

Fig. 1. FOD on simulated data with SNR = 2: (a) simulated fiber directions; (b) baseline method; (c) isotropic and (d) anisotropic spatial regularization.

neighboring voxels. The fiber direction R_{angle} of each voxel in Fig. 1(A) is :

$$R_{angle} = (i - 1)\theta + j\theta, \tag{9}$$

where i is row number, j is column number. We used $\theta \in \{10°, 20°, 30°, 40°\}$ to generate four sets of simulated data. Rician noise was added to generate data with SNR $= 2$. In the second experiment, we used the pattern in Fig. 1(B) to compare the performance of FOD reconstruction algorithm for resolving fiber crossings. By varying the noise level, we also generated four sets of simulated data with SNR=2,3,4,5. For each voxel, the angular error of each fiber direction on the unit sphere is computed as its angle to the nearest true fiber direction. The overall AAE at each voxel is then a weighted average of angular errors from all directions with the FOD value as the weight. To measure the reconstruction accuracy, we computed the *average angular error* (AAE) for all the 121 voxels in Fig. 1(A) or (E).

In Fig. 1(B), the AAE measures of the three FOD reconstruction methods for the first fiber pattern are shown. We can clearly see that the whole-brain reconstruction method with anisotropic diffusion achieved the best performance in all four sets of simulation data. For data simulated with $\theta = 10°$, reconstruction results from the baseline and whole-brain method with anisotropic regularization are shown in Fig. 1(C) and (D), respectively, where we can see the latter genereates much cleaner FOD reconstruction without spurious fiber directions as compared to the true fiber pattern in Fig. 1(A). For the second experiment, the overall AAE mesures from the three algorithms are presented under different SNR levels. We can see that both whole-brain reconstruction methods outperformed the voxel-wise method, especially at low SNR levels. Whole-brain reconstruction methods with isotroic and anisotropic regularization achieved similar level of performance in this experiment. For data with SNR=2, reconstruction results from the baseline and the whole-brain method with anisotropic regularization are plotted in Fig. 1(G) and (H), respectively. Clearly the whole-brain method produces much more accurate reconstruction of the fiber pattern shown in Fig. 1(E). Considering the overall performance of these two whole-brain reconstruction methods, we will choose the anisotropic regularation in our large-scale experiments with HCP data.

3.2   HCP Data for Locus Coeruleus Bundle Reconstruction

In this experiment, we will apply both voxel-wise and whole-brain FOD reconstruction to 80 subjects from HCP and evaluate their impact on the reconstruction of challenging bundles from the locus coeruleus (LC) to amygadala. This is a critical problem because the LC nuclei in the brain stem is considered the earliest location with tau pathology in the latest Braak staging. It is receiving increasing interesting interests in aging and Alzheimer's disease (AD) research. For studying transynaptic propoation of tau tangles from LC to medial temporal lobe, it is thus critical to robustly reconstruct the LC fiber bundles. For voxel-wise FOD reconstruction, we apply the baseline method without spatial

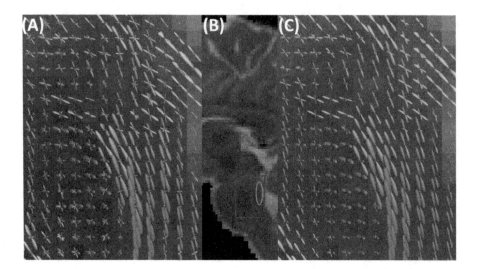

Fig. 2. A comparison of FOD reconstruction results in a brain stem ROI of an HCP subject. The brainstem ROI (red rectangle) is shown in (B), where the LC nuclei is highlighted with the green ellipse. Reconstructed FODs with the voxel-wise and whole-brain approach are shown in (A) and (C), respectively. (Color figure online)

regularization [5]. For whole-brain FOD reconstruction, we apply anisotropic diffusion in the operator splitting algorithm because it achieved overall the most robust performance in our simulations. For LC bundle reconstruction with tractography, we nonlinear warped an LC atlas to generate the seed region in each HCP subject and the amygdala segmentation from HCP as the arget ROI.

As a demonstration, we show in Fig. 2 the FOD reconstruction results in a brainstem ROI from both methods. Clearly we can see the whole-brain reconstruction method generates a more regular field of FODs with much less spurious fiber directions. To test the effect of the FODs on fiber bundle reconstruction, we apply the exact same parameters and ROIs to run FOD-based probabilistic tractography from the MRTrix software [13] from the LC seed region to the amygdala. The specific parameters are as follows: curvature=1.5; FOD threshold=0.05; maximum number of tracts = 200K. Results from four representative subjects are shown in Fig. 3. We can see that the fiber bundles based on wholebrain FOD reconstruction provide a more robust representation of the LC to amygdala bundle as indicated by the larger number of tracts and smoother trajectories. In Fig. 4, we plotted the total number of fiber tracts constained in the LC bundles reconstructed from both methods. We can see that the same trend holds for all subjects used in our experiment. While tract number is not a biologically meaningful indicator, these results show that whole-brain FOD reconstruction can improve the robustness of fiber bundle reconstruction, especially for challenging cases such as the LC fiber bundle.

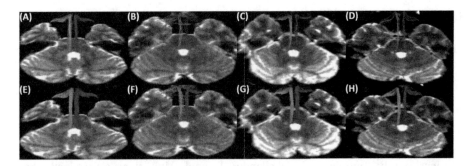

Fig. 3. LC bundles from four HCP subjects (left: red; right: blue). For each subject, the bundles computed from voxel-wise and whole-brain reconstruction methods are plotted on the top and bottome row of each column, respectively. (A)-(D): LC bundles based on FODs from voxel-wise reconstruction; (E)-(H): LC bundles from whole-brain FOD reconstruction methods. (Color figure online)

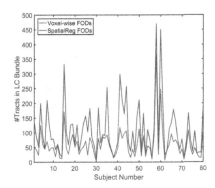

Fig. 4. The number of fiber tracts in the LC bundle reconstructed from FODs based on voxel-wise and whole-brain reconstruction methods.

4   Discussion and Conclusion

In this work, we develop a novel computational framework for whole-brain FOD reconstruction from multi-shell imaging data. Promising experimental results from numerical simulations and large-scale HCP data have been presented. In particular, we applied the proposed method to large-scale HCP data for the reconstruction of the challenging fiber bundles from the LC nuclei in the brainstem, which are emerging as a critical area for AD research. We demonstrate that more robust fiber bundle reconstruction can be achieved with the proposed whole-brain approach for FOD reconstruction. To the best of our knowledge, this is the first large-scale evaluation of FOD reconstruction with spatial regularization. For future work, we will further validate our novel methods on the reconstruction of challenging fiber bundles and graph-based connectivity analysis.

References

1. Essen, D.V., Ugurbil, K., et al.: The human connectome project: a data acquisition perspective. NeuroImage **62**(4), 2222–2231 (2012)
2. Ugurbil, K., Xu, J., Auerbach, E.J., et al.: Pushing spatial and temporal resolution for functional and diffusion MRI in the human connectome project. NeuroImage **80**, 80–104 (2013)
3. Panagiotaki, E., Schneider, T., Siow, B., et al.: Compartment models of the diffusion MR signal in brain white matter: a taxonomy and comparison. Neuroimage **59**(3), 2241–2254 (2012)
4. Jeurissen, B., Tournier, J.D., Dhollander, T., et al.: Multi-tissue constrained spherical deconvolution for improved analysis of multi-shell diffusion MRI data. NeuroImage **103**(0), 411–426 (2014)
5. Tran, G., Shi, Y.: Fiber orientation and compartment parameter estimation from multi-shell diffusion imaging. IEEE T. Med. Imaging **34**(11), 2320–2332 (2015)
6. Laidlaw, D.H., Weickert, J.: Visualization and Processing of Tensor Fields. Advances and perspectives. Springer Science & Business Media, Berlin (2009)
7. Li, J., Shi, Y., Toga, A.W.: Diffusion of fiber orientation distribution functions with a rotation-induced riemannian metric. In: Golland, P., Hata, N., Barillot, C., Hornegger, J., Howe, R. (eds.) MICCAI 2014. LNCS, vol. 8675, pp. 249–256. Springer, Cham (2014). doi:10.1007/978-3-319-10443-0_32
8. Goh, A., Lenglet, C., Thompson, P.M., Vidal, R.: Estimating orientation distribution functions with probability density constraints and spatial regularity. In: Yang, G.-Z., Hawkes, D., Rueckert, D., Noble, A., Taylor, C. (eds.) MICCAI 2009. LNCS, vol. 5761, pp. 877–885. Springer, Heidelberg (2009). doi:10.1007/978-3-642-04268-3_108
9. Zhou, Q., Michailovich, O., Rathi, Y.: Spatially regularized reconstruction of fibre orientation distributions in the presence of isotropic diffusion. arXiv preprint arXiv:1401.6196 (2014)
10. Reisert, M., Kiselev, V.G.: Fiber continuity: an anisotropic prior for ODF estimation. IEEE T. Med. Imaging **30**(6), 1274–1283 (2011)
11. Zhang, X., Burger, M., Bresson, X., et al.: Bregmanized nonlocal regularization for deconvolution and sparse reconstruction. SIAM J. Img. Sci. **3**(3), 253–276 (2010)
12. Braak, H., Thal, D.R., Ghebremedhin, E., Del Tredici, K.: Stages of the pathologic process in alzheimer disease: age categories from 1 to 100 years. J. Neuropathol. Exp. Neurol. **70**(11), 960 (2011)
13. Tournier, J.D., Calamante, F., Connelly, A.: Mrtrix: diffusion tractography in crossing fiber regions. Int. J. Imag. Syst. Tech. **22**(1), 53–66 (2012)

Topological Distances Between Brain Networks

Moo K. Chung[1(✉)], Hyekyoung Lee[2], Victor Solo[3], Richard J. Davidson[1],
and Seth D. Pollak[1]

[1] University of Wisconsin, Madison, USA
mkchung@wisc.edu
[2] Seoul National University, Seoul, Korea
[3] University of New South Wales, Sydney, Australia

Abstract. Many existing brain network distances are based on matrix
norms. The element-wise differences may fail to capture underlying topological differences. Further, matrix norms are sensitive to outliers. A few
extreme edge weights may severely affect the distance. Thus it is necessary to develop network distances that recognize topology. In this paper,
we introduce Gromov-Hausdorff (GH) and Kolmogorov-Smirnov (KS)
distances. GH-distance is often used in persistent homology based brain
network models. The superior performance of KS-distance is contrasted
against matrix norms and GH-distance in random network simulations
with the ground truths. The KS-distance is then applied in characterizing
the multimodal MRI and DTI study of maltreated children.

1   Introduction

There are many similarity measures and distances between networks in literature
[2,7,14]. Many of these approaches simply ignore the topology of the networks
and mainly use the sum of differences between either node or edge measurements.
These network distances are sensitive to the topology of networks. They may
lose sensitivity over topological structures such as the connected components,
modules and holes in networks.

In standard graph theoretic approaches, the similarity and distance of networks are measured by determining the difference in graph theory features such
as assortativity, betweenness centrality, small-worldness and network homogene-
ity [4,17]. Comparison of graph theory features appears to reveal changes of
structural or functional connectivity associated with different clinical populations [17]. Since weighted brain networks are difficult to interpret and visualize,
they are often turned into binary networks by thresholding edge weights [11,20].
However, the choice of thresholding the edge weights may alter the network
topology. To obtain the proper optimal threshold, the multiple comparison correction over every possible edge has been proposed [16,18,20]. However, depending on what p-value to threshold, the resulting binary graph also changes. Others
tried to control the sparsity of edges in the network in obtaining the binary network [11,20]. However, one encounters the problem of thresholding sparse parameters. Thus existing methods for binarizing weighted networks cannot escape
the inherent problem of arbitrary thresholding.

© Springer International Publishing AG 2017
G. Wu et al. (Eds.): CNI 2017, LNCS 10511, pp. 161–170, 2017.
DOI: 10.1007/978-3-319-67159-8_19

Until now, there is no widely accepted criteria for thresholding networks. Instead of trying to come up with an optimal threshold for network construction that may not work for different clinical populations or cognitive conditions [20], *why not use all networks for every possible threshold?* Motivated by this question, new multiscale hierarchical network modeling framework based on persistent homology has been developed recently [7,14]. In persistent homology based brain network analysis as first formulated in [14], we build the collection of nested networks over every possible threshold using the *graph filtration*, a persistent homological construct [14]. The graph filtration is a threshold-free framework for analyzing a family of graphs but requires hierarchically building specific nested subgraph structures. The graph filtration shares similarities to the existing multi-thresholding or multi-resolution network models that use many different arbitrary thresholds or scales [11,14]. Such approaches are mainly used to visually display the dynamic pattern of how graph theoretic features change over different thresholds and the pattern of change is rarely quantified. Persistent homology can be used to quantify such dynamic pattern in a more coherent mathematical framework.

In persistent homology, there are various metrics that have been proposed to measure network distance. Among them, *Gromov-Hausdorff (GH) distance* is possibly the most popular distance that is originally used to measure distance between two metric spaces [19]. It was later adapted to measure distances in persistent homology, dendrograms [5] and brain networks [14]. The probability distributions of GH-distance is unknown. Thus, the statistical inference on GH-distance has been done through resampling techniques such as jackknife, bootstraps or permutations [7,14,15], which often cause computational bottlenecks for large-scale networks. To bypass the computational bottleneck associated with resampling large-scale networks, the *Kolmogorov-Smirnov (KS) distance* was introduced in [6,8,15]. The advantage of using KS-distance is its easiness to interpret compared to other less intuitive distances from persistent homology. Due to its simplicity, it is possible to determine its probability distribution exactly [8].

Many distance or similarity measures are not metrics but having metric distances makes the interpretation of brain networks easier due to the triangle inequality. Further, existing network distance concepts are often borrowed from the metric space theory. Let us start with formulating networks as metric spaces.

2   Matrix Norms

Consider a weighted graph or network with the node set $V = \{1,\ldots,p\}$ and the edge weights $w = (w_{ij})$, where w_{ij} is the weight between nodes i and j. We may assume that the edge weights satisfy the metric properties: nonnegativity, identity, symmetry and the triangle inequality such that

$$w_{i,j} \geq 0, \; w_{ii} = 0, \; w_{ij} = w_{ji}, \; w_{ij} \leq w_{ik} + w_{kj}.$$

With theses conditions, $\mathcal{X} = (V, w)$ forms a metric space. Although the metric property is not necessary for building a network, it offers many nice mathematical properties and easier interpretation on network connectivity.

Example 1. Given measurement vector $\mathbf{x}_i = (x_{1i}, \cdots, x_{ni})^\top \in \mathbb{R}^n$ on the node i. The weight $w = (w_{ij})$ between nodes is often given by some bivariate function f: $w_{ij} = f(\mathbf{x}_i, \mathbf{x}_j)$. The correlation between $\mathbf{x}_i$ and $\mathbf{x}_j$, denoted as $\mathrm{corr}(\mathbf{x}_i, \mathbf{x}_j)$, is a bivariate function. If the weights $w = (w_{ij})$ are given by $w_{ij} = \sqrt{1 - \mathrm{corr}(\mathbf{x}_i, \mathbf{x}_j)}$, it can be shown that $\mathcal{X} = (V, w)$ forms a metric space.

Matrix norm of the difference between networks is often used as a measure of similarity between networks [2,21]. Given two networks $\mathcal{X}^1 = (V, w^1)$ and $\mathcal{X}^2 = (V, w^2)$, the L_l-norm of network difference is given by

$$D_l(\mathcal{X}^1, \mathcal{X}^2) = \| w^1 - w^2 \|_l = \left(\sum_{i,j} |w_{ij}^1 - w_{ij}^2|^l \right)^{1/l}.$$

Note L_l is the element-wise Euclidean distance in l-dimension. When $l = \infty$, L_∞-distance is written as

$$D_\infty(\mathcal{X}^1, \mathcal{X}^2) = \| w^1 - w^2 \|_\infty = \max_{\forall i,j} |w_{ij}^1 - w_{ij}^2|.$$

The element-wise differences may not capture additional higher order similarity. For instance, there might be relations between a pair of columns or rows [21]. Also L_1 and L_2-distances usually surfer the problem of outliers. Few outlying extreme edge weights may severely affect the distance. Further, these distances ignore the underlying topological structures. Thus, there is a need to define distances that are more topological.

3   Gromov-Hausdorff Distance

GH-distance for brain networks is first introduced in [14]. GH-distance measures the difference between networks by embedding the network into the ultrametric space that represents hierarchical clustering structure of network [5]. The distance s_{ij} between the closest nodes in the two disjoint connected components $\mathbf{R}_1$ and $\mathbf{R}_2$ is called the single linkage distance (SLD), which is defined as

$$s_{ij} = \min_{l \in \mathbf{R}_1, k \in \mathbf{R}_2} w_{lk}.$$

Every edge connecting a node in $\mathbf{R}_1$ to a node in $\mathbf{R}_2$ has the same SLD. SLD is then used to construct the single linkage matrix (SLM) $S = (s_{ij})$ (Fig. 1). SLM shows how connected components are merged locally and can be used in constructing a dendrogram. SLM is a *ultrametric* which is a metric space satisfying the stronger triangle inequality $s_{ij} \leq \max(s_{ik}, s_{kj})$ [5]. Thus the dendrogram can be represented as a ultrametric space $\mathcal{D} = (V, S)$, which is again a metric space. GH-distance between networks is then defined through GH-distance

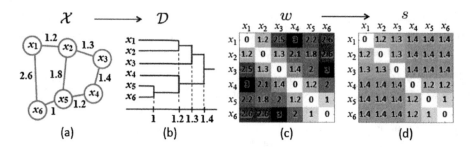

Fig. 1. (a) Toy network, (b) its dendrogram, (c) the distance matrix w based on Euclidean distance, (d) the single linkage matrix (SLM) S.

between corresponding dendrograms. Given two dendrograms $\mathcal{D}^1 = (V, S^1)$ and $\mathcal{D}^2 = (V, S^2)$ with SLM $S^1 = (s_{ij}^1)$ and $S^2 = (s_{ij}^2)$,

$$D_{GH}(\mathcal{D}^1, \mathcal{D}^2) = \frac{1}{2} \max_{\forall i,j} |s_{ij}^1 - s_{ij}^2|. \tag{1}$$

For the statistical inference on GH-distance, resampling techniques such as jackknife or permutation tests are often used [14,15].

4   Kolmogorov-Smirnov Distance

Recently a new network distance based on the concept of graph filtration has been proposed in [8]. Given weighted network $\mathcal{X} = (V, w)$, the binary network $\mathcal{B}_\epsilon(\mathcal{X}) = (V, \mathcal{B}_\epsilon(w))$ is a graph consisting of the node set V and the edge weight $\mathcal{B}_\epsilon(w) = (\mathcal{B}_\epsilon(w_{ij}))$ given by

$$\mathcal{B}_\epsilon(w_{ij}) = \begin{cases} 1 & \text{if } w_{ij} \leq \epsilon; \\ 0 & \text{otherwise.} \end{cases} \tag{2}$$

Note $\mathcal{B}_\epsilon(w)$ is the adjacency matrix of $\mathcal{B}_\epsilon(\mathcal{X})$. Then it can be shown that

$$\mathcal{B}_{\epsilon_0}(\mathcal{X}) \subset \mathcal{B}_{\epsilon_1}(\mathcal{X}) \subset \mathcal{B}_{\epsilon_2}(\mathcal{X}) \subset \cdots$$

for $0 = \epsilon_0 \leq \epsilon_1 \leq \epsilon_2 \cdots$. The sequence of such nested multiscale graph structure is called the *graph filtration* [7,14]. The sequence of thresholded values $\epsilon_0, \epsilon_1, \epsilon_2 \cdots$ are called the *filtration values*.

The graph filtration can be quantified using monotonic function f satisfying

$$f \circ \mathcal{B}_{\epsilon_j}(\mathcal{X}) \geq f \circ \mathcal{B}_{\epsilon_{j+1}}(\mathcal{X})$$

for $\epsilon_j \leq \epsilon_{j+1}$. The number of connected components, the zeroth Betti number β_0, satisfies the monotonicity property (3). The size of the largest cluster, denoted as γ, satisfies a similar but opposite relation of monotonic increase [7].

Given two networks $\mathcal{X}^1 = (V, w^1)$ and $\mathcal{X}^2 = (V, w^2)$, Kolmogorov-Smirnov (KS) distance between $\mathcal{X}^1$ and $\mathcal{X}^2$ is defined as [7,15]

$$D_{KS}(\mathcal{X}^1, \mathcal{X}^2) = \sup_{\epsilon \geq 0} \left| f \circ \mathcal{B}_\epsilon(\mathcal{X}^1) - f \circ \mathcal{B}_\epsilon(\mathcal{X}^2) \right|.$$

The distance D_{KS} is motivated by Kolmogorov-Smirnov (KS) test for determining the equivalence of two cumulative distribution functions [8,10].

Example 2. Consider network with edge weights $r_{ij} = 1 - \text{corr}(\mathbf{x}_i, \mathbf{x}_j)$. Such network is not a metric space. To make it a metric space, we need to scale the edge weight to $w_{ij} = \sqrt{r_{ij}}$ (Example 1). However, KS-distance is invariant under such monotonic scaling since the distance is taken over every possible filtration value.

The distance D_{KS} can be discretely approximated using the finite number of filtrations:

$$D_q = \sup_{1 \leq j \leq q} \left| f \circ \mathcal{B}_{\epsilon_j}(\mathcal{X}^1) - f \circ \mathcal{B}_{\epsilon_j}(\mathcal{X}^2) \right|.$$

If we choose enough number of q such that ϵ_j are all the sorted edge weights, then $D_{KS}(\mathcal{X}^1, \mathcal{X}^2) = D_q$ [8]. This is possible since there are only up to $p(p-1)/2$ number of unique edges in a graph with p nodes and $f \circ \mathcal{B}_\epsilon$ increases discretely. In practice, ϵ_j may be chosen uniformly.

The probability distribution of D_q under the null is asymptotically given by

$$\lim_{q \to \infty} \left(D_q / \sqrt{2q} \geq d \right) = 2 \sum_{i=1}^{\infty} (-1)^{i-1} e^{-2i^2 d^2}. \tag{3}$$

The result is first given in [8]. p-value under the null is then computed as

$$p\text{-value} = 2e^{-d_o^2} - 2e^{-8d_o^2} + 2e^{-18d_o^2} \cdots,$$

where the observed value d_o is the least integer greater than $D_q / \sqrt{2q}$ in the data. For any large value $d_0 \geq 2$, the second term is in the order of 10^{-14} and insignificant. Even for small observed d_0, the expansion converges quickly and 5 terms are sufficient. KS-distance method does not assume any statistical distribution on graph features other than that they has to be monotonic. The technique is very general and applicable to other monotonic graph features such as node degrees.

5   Comparisons

Five different network distances (L_1, L_2, L_∞, GH and KS) were compared in simulation studies with modular structures. The simulations below were independently performed 100 times and the average results were reported.

There were four groups and the sample size was $n = 5$ in each group and the number of nodes was $p = 100$ (Fig. 2). We follow notations in Example 1.

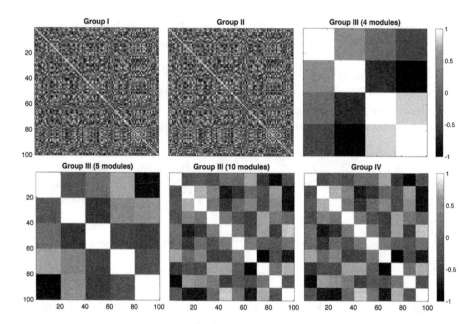

Fig. 2. Randomly simulated correlation matrices. Group I and Group II were generated independently and identically. Group III was generated from Group I but additional dependency was added to introduce modular structures. Group IV was generated from Group III (10 modules) by adding small noise.

In Group I, the measurement vector $\mathbf{x}_i$ at node i was simulated as multivariate normal, i.e., $\mathbf{x}_i \sim N(0, I_n)$ with n by n identity matrix I_n as the covariance matrix. The edge weights for group I was $w_{ij}^1 = \sqrt{1 - \mathrm{corr}(\mathbf{x}_i, \mathbf{x}_j)}$. In Group II, the measurement vector $\mathbf{y}_i$ at node i was simulated as $\mathbf{y}_i = \mathbf{x}_i + N(0, \sigma^2 I_n)$ with noise level $\sigma = 0.01$. The edge weight for group II was $w_{ij}^2 = \sqrt{1 - \mathrm{corr}(\mathbf{y}_i, \mathbf{y}_j)}$.

Group III was generated by adding additional dependency to Group I:

$$\mathbf{y}_i = 0.5\mathbf{x}_{ci+1} + N(0, \sigma I_n).$$

This introduce modules in the network. We assumed there were total $k = 4, 5, 10$ modules and each module consists of $c = p/k$ number of points. Group IV was generated by adding noise to Group III: $\mathbf{z}_i = \mathbf{y}_i + N(0, \sigma^2 I_n)$.

No network difference. It was expected there was no network difference between Groups I and II. We applied the 5 different distances. For the first four distances, permutation test was used. Since there were 5 samples in each group, the total number of permutations was $\binom{10}{5} = 272$ making the permutation test exact and the comparisons fair. All the distances performed well and did not detect network differences (1st row in Table 1). It was also expected there is no network difference between Groups III and IV. We compared 4 module network to 4 module network. All the distances performed equally well and did not detect differences (2nd row in Table 1).

Table 1. Simulation results given in terms of p-values. In the case of no network differences (0 vs. 0 and 4 vs. 4), higher p-values are better. In the case of network differences (4 vs. 5 and 5 vs. 10), smaller p-values are better. * and ** indicates multiplying 10^{-3} and 10^{-4}.

	L_1	L_2	L_∞	GH	KS (β_0)	KS (γ)
0 vs. 0	0.93 ± 0.04	0.93 ± 0.04	0.93 ± 0.04	0.87 ± 0.14	1.00 ± 0.00	1.00 ± 0.00
4 vs. 4	0.89 ± 0.02	0.89 ± 0.02	0.90 ± 0.03	0.86 ± 0.17	0.87 ± 0.29	0.88 ± 0.28
4 vs. 5	0.14 ± 0.16	0.06 ± 0.10	0.03 ± 0.06	0.29 ± 0.30	$(0.07 \pm 0.67)^{**}$	$(0.07 \pm 0.67)^{**}$
5 vs. 10	0.47 ± 0.25	0.19 ± 0.18	0.10 ± 0.10	0.33 ± 0.30	0.01 ± 0.08	$(0.06 \pm 0.53)^{*}$

Network difference. Networks with 4, 5 and 10 modules were generated using Group III models. Since the number of modules were different, they were considered as different networks. We compared 4 and 5 module networks (3rd row in Table 1), and 5 and 10 module networks (4th row in Table 1). L_1, L_2, L_∞ distances did not performed well for 5 vs. 10 module comparisons. Surprisingly, GH-distance performed worse than L_∞ in all cases. On the other hand, KS-distance performed extremely well.

The results of the above simulations did not change much even if we increased the noise level to $\sigma = 0.1$. In terms of computation, distance methods based on the permutation test took about 950 s (16 min) while the KS-like test procedure only took about 20 s in a computer. The MATLAB code for performing these simulations is given in http://www.cs.wisc.edu/~mchung/twins. The results given in Table 1 may slightly change if different random networks are generated.

6   Application

The methods were applied to multimodal MRI and DTI of 31 normal controls and 23 age-matched children who experienced maltreatment while living in post-institutional settings before being adopted by families in US. The detailed deception of the subject and image acquisition parameters are given in [7]. Ages range from 9 to 14 years. The average amount of time spend in institutional care was 2.5 ± 1.4 years. Children were on average 3.2 years when they were adapted.

For MRI, a study specific template was constructed using the diffeomorphic shape and intensity averaging technique through Advanced Normalization Tools (ANTS) [1]. White matter was also segmented into tissue probability maps using template-based priors, and registered to the template [3]. The Jacobian determinants of the inverse deformations from the template to individual subjects were obtained. DTI were corrected for eddy current related distortion and head motion via FSL (http://www.fmrib.ox.ac.uk/fsl) and distortions from field inhomogeneities were corrected [12] before performing a non-linear tensor estimation using CAMINO [9]. Subsequently, iterative tensor image registration strategy was used for spatial normalization [13]. Then fractional anisotropy (FA) were

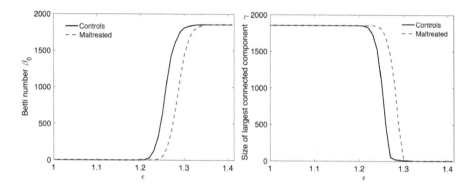

Fig. 3. The plots of β_0 (left) and γ (right) over $\sqrt{1 - \text{corr.}}$ showing structural network differences between maltreated children (dotted red) and normal controls (solid black) on 1856 nodes. (Color figure online)

calculated for diffusion tensor volumes diffeomorphically registered to the study specific template. Jacobian determinants and FA-values are uniformly sampled at 1856 nodes along the white mater template boundary.

Correlation within modality. The correlations of the Jacobian determinant and FA-values were computed between nodes within each modality. This results in 1856×1856 correlation matrix for each group and modality. Using KS-distance, we determined the statistical significance of the correlation matrix differences between the groups for each modality separately. The statistical results in terms of p-values are all below 0.0001 indicating the very strong overall structural network differences in both MRI and DTI.

Cross-correlation across modality. Following the hyper-network framework in [8], we also computed the cross-correlation between the Jacobian determinants and FA-values on 1856 nodes. This results in 1856×1856 cross-correlation matrix for each group. The statistical significance of the cross-correlation matrix differences is then determined using KS-distance (p-value < 0.0001). The KS-distance method is robust under the change of node size and we also obtained the similar result when the node size changed to 548.

7   Discussion

The limitation of GH- and KS-distances. The limitation of the SLM is the inability to discriminate a cycle in a graph. Consider two topologically different graphs with three nodes (Fig. 4). However, the corresponding SLM are identically given by

$$\begin{pmatrix} 0 & 0.2 & 0.5 \\ 0.2 & 0 & 0.5 \\ 0.5 & 0.5 & 0 \end{pmatrix} \text{ and } \begin{pmatrix} 0 & 0.2 & 0.5 \\ 0.2 & 0 & 0.5 \\ 0.5 & 0.5 & 0 \end{pmatrix}.$$

The lack of uniqueness of SLMs makes GH-distance incapable of discriminating networks with cycles [6]. KS-distance also treat the two networks in Fig. 4 as identical if Betti number β_0 is used as the monotonic feature function. Thus, KS-distance also fail to discriminate cycles.

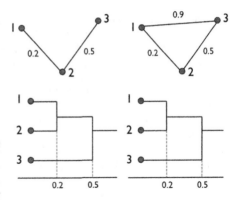

Computation. The total number of permutations in permuting two groups of size q each is [8] $\binom{2q}{q} \sim \frac{4^q}{\sqrt{2\pi q}}$. Even for small $q = 10$, more than tens of thousands permutations are needed for the accurate estimation the p-value. On the other hand, only up to 10 terms are needed in the KS-distance method.

Fig. 4. Two topologically distinct graphs may have identical dendrograms, which results in zero GH-distance.

The KS-distance method avoids the computational burden of permutation tests.

Acknowledgements. This work is supported by NIH Grants MH61285, MH68858, MH84051, UL1TR000427, Brain Initiative Grant EB022856 and Basic Science Research Program through the National Research Foundation (NRF) of Korea (NRF-2016R1D1A1B03935463). M.K.C. would like to thank professor A.M. Mathai of McGill University for asking to prove the convergence of KS test in a homework. That homework motivated the construction of KS-distance for graphs.

References

1. Avants, B.B., Epstein, C.L., Grossman, M., Gee, J.C.: Symmetric diffeomorphic image registration with cross-correlation: evaluating automated labeling of elderly and neurodegenerative brain. Med. Image Anal. **12**, 26–41 (2008)
2. Banks, D., Carley, K.: Metric inference for social networks. J. Classif. **11**, 121–149 (1994)
3. Bonner, M.F., Grossman, M.: Gray matter density of auditory association cortex relates to knowledge of sound concepts in primary progressive aphasia. J. Neurosci. **32**, 7986–7991 (2012)
4. Bullmore, E., Sporns, O.: Complex brain networks: graph theoretical analysis of structural and functional systems. Nat. Rev. Neurosci. **10**, 186–98 (2009)
5. Carlsson, G., Mémoli, F.: Characterization, stability and convergence of hierarchical clustering methods. J. Mach. Learn. Res. **11**, 1425–1470 (2010)
6. Chung, M.K.: Computational Neuroanatomy: The Methods. World Scientific, Singapore (2012)
7. Chung, M.K., Hanson, J.L., Ye, J., Davidson, R.J., Pollak, S.D.: Persistent homology in sparse regression and its application to brain morphometry. IEEE Trans. Med. Imaging **34**, 1928–1939 (2015)

8. Chung, M.K., Villalta-Gil, V., Lee, H., Rathouz, P.J., Lahey, B.B., Zald, D.H.: Exact topological inference for paired brain networks via persistent homology. In: Niethammer, M., Styner, M., Aylward, S., Zhu, H., Oguz, I., Yap, P.-T., Shen, D. (eds.) IPMI 2017. LNCS, vol. 10265, pp. 299–310. Springer, Cham (2017). doi:10.1007/978-3-319-59050-9_24

9. Cook, P.A., Bai, Y., Nedjati-Gilani, S., Seunarine, K.K., Hall, M.G., Parker, G.J., Alexander, D.C.: Camino: open-source diffusion-MRI reconstruction and processing. In: 14th Scientific Meeting of the International Society for Magnetic Resonance in Medicine (2006)

10. Gibbons, J.D., Chakraborti, S.: Nonparametric Statistical Inference. Chapman & Hall/CRC Press, Boca Raton (2011)

11. He, Y., Chen, Z., Evans, A.: Structural insights into aberrant topological patterns of large-scale cortical networks in Alzheimer's disease. J. Neurosci. **28**, 4756 (2008)

12. Jezzard, P., Clare, S.: Sources of distortion in functional MRI data. Hum. Brain Mapp. **8**, 80–85 (1999)

13. Joshi, S.C., Davis, B., Jomier, M., Gerig, G.: Unbiased diffeomorphic atlas construction for computational anatomy. NeuroImage **23**, 151–160 (2004)

14. Lee, H., Kang, H., Chung, M.K., Kim, B.-N., Lee, D.S.: Persistent brain network homology from the perspective of dendrogram. IEEE Trans. Med. Imaging **31**, 2267–2277 (2012)

15. Lee, H., Kang, H., Chung, M.K., Lim, S., Kim, B.-N., Lee, D.S.: Integrated multimodal network approach to PET and MRI based on multidimensional persistent homology. Hum. Brain Mapp. **38**, 1387–1402 (2017)

16. Rubinov, M., Knock, S. A., Stam, C. J., Micheloyannis, S., Harris, A.W., Williams, L.M., Breakspear, M.: Small-world properties of nonlinear brain activity in schizophrenia

17. Rubinov, M., Sporns, O.: Complex network measures of brain connectivity: uses and interpretations. NeuroImage **52**, 1059–1069 (2010)

18. Salvador, R., Suckling, J., Coleman, M.R., Pickard, J.D., Menon, D., Bullmore, E.: Neurophysiological architecture of functional magnetic resonance images of human brain. Cereb. Cortex **15**, 1332–1342 (2005)

19. Tuzhilin, A.A.: Who invented the Gromov-Hausdorff distance? arXiv preprint arXiv:1612.00728 (2016)

20. Wijk, B.C.M., Stam, C.J., Daffertshofer, A.: Comparing brain networks of different size and connectivity density using graph theory. PLoS ONE **5**, e13701 (2010)

21. Zhu, X., Suk, H.-I., Shen, D.: Matrix-similarity based loss function and feature selection for alzheimer's disease diagnosis. In: Proceedings of the IEEE Conference on Computer Vision and Pattern Recognition, pp. 3089–3096 (2014)

Author Index

Adeli, Ehsan 17
Al-Arif, S.M. Masudur Rahman 70
Alonso, Eduardo 70
Anderson, Jeffrey 98
Asad, Muhammad 70

Bonilha, Leonardo 79
Bouix, Sylvain 108
Breier, Alan 108

Chamberland, Maxime 35
Chen, Xiaobo 9
Chung, Moo K. 134, 161
Corr, Philip 70
Crimi, Alessandro 1

Davidson, Richard J. 134, 161
Delgaizo, John 79
Dentico, Daniela 134
Descoteaux, Maxime 35
Dima, Danai 70

Fletcher, P. Thomas 98

Gray, William 35

Hofesmann, Eric 79

Jia, Xiuyi 17
Jones, Derek K. 35
Jose, Vipin 98
Joshi, Shantanu H. 125
Jun, Eunji 143

Leaver, Amber M. 125
Lee, David S. 125
Lee, Hyekyoung 161
Lei, Du 25
Li, Junling 152
Lin, Weili 116
Lisowska, Anna 42
Lutz, Antoine 134

Mechelli, Andrea 25
Mohamed, Saida S. 1
Mostofsky, Stewart 60
Munsell, Brent C. 79, 89

Narr, Katherine L. 125
Nebel, Mary Beth 60
Nesi, Lucas L. 89
Nguyen, Nancy Duong 1

Palande, Sourabh 98
Pasternak, Ofer 108
Pollak, Seth D. 161

Rekik, Islem 42, 51
Riaz, Atif 70
Rorden, Chris 89

Shen, Dinggang 9, 17, 116
Shenton, Martha E. 108
Shi, Yonggang 152
Slabaugh, Greg 70
Solo, Victor 161
Soussia, Mayssa 51
Styner, Martin 79
Suk, Heung-Il 143
Sun, Wei 152
Swago, Sophia 108

Venkataraman, Archana 60

Wang, Bei 98
Wang, Yuan 134
West, John D. 108
Woods, Roger P. 125
Wymbs, Nicholas 60

Yin, Weiyan 116
Yoneki, Eiko 1
Young, Jonathan 25

Zhang, Han 9, 17, 116
Zhang, Yu 9
Zielinski, Brandon 98

Printed in the United States
By Bookmasters